PHYSIOLOGICAL DIVERSITY AND
ITS ECOLOGICAL IMPLICATIONS

Physiological Diversity and its Ecological Implications

JOHN I. SPICER
Lecturer in Zoology

KEVIN J. GASTON
Royal Society University Research Fellow

Department of Animal & Plant Sciences
University of Sheffield

**Blackwell
Science**

© 1999 by
Blackwell Science Ltd
Editorial Offices:
Osney Mead, Oxford OX2 0EL
25 John Street, London WC1N 2BL
23 Ainslie Place, Edinburgh
  EH3 6AJ
350 Main Street, Malden
  MA 02148-5018, USA
54 University Street, Carlton
  Victoria 3053, Australia
10, rue Casimir Delavigne
  75006 Paris, France

Other Editorial Offices:
Blackwell Wissenschafts-Verlag
  GmbH
Kurfürstendamm 57
10707 Berlin, Germany

Blackwell Science KK
MG Kodenmacho Building
7–10 Kodenmacho Nihombashi
Chuo-ku, Tokyo 104, Japan

First published 1999

Set by Excel Typesetters Co., Hong
Kong
Printed and bound in Great Britain
at the Alden Press Ltd, Oxford and
Northampton

A catalogue record for this title
is available from the British Library

ISBN 0-632-05452-2

Library of Congress
Cataloging-in-publication Data

Spicer, John I.
    Physiological diversity and its
    ecological implications/John
  I. Spicer, Kevin J. Gaston.
        p.   cm.
    ISBN 0-632-05452-2
    1. Ecophysiology.   I.
    Gaston, Kevin J.   II. Title.
QH541. 15. E26S68   1999
    591.7'8—dc21   99-16986
                              CIP

DISTRIBUTORS

  Marston Book Services Ltd
  PO Box 269
  Abingdon, Oxon OX14 4YN
  (*Orders*: Tel: 01235 465500
          Fax: 01235 465555)

USA
  Blackwell Science, Inc.
  Commerce Place
  350 Main Street
  Malden, MA 02148-5018
  (*Orders*: Tel: 800 759 6102
              781 388 8250
          Fax: 781 388 8255)

Canada
  Login Brothers Book Company
  324 Saulteaux Crescent
  Winnipeg, Manitoba R3J 3T2
  (*Orders*: Tel: 204 837 2987)

Australia
  Blackwell Science Pty Ltd
  54 University Street
  Carlton, Victoria 3053
  (*Orders*: Tel: 3 9347 0300
          Fax: 3 9347 5001)

For further information on
Blackwell Science, visit our website:
www.blackwell-science.com

Contents

References, 195

Index, 235

Preface

It is but a short step from where some of this book was written to the pier which protrudes into the bay at Millport, on the Isle of Cumbrae off the west coast of Scotland. The events unfolding beneath that pier one afternoon in July serve to exemplify some of the issues which the book addresses. Peering through a thick film of oil on the surface of the water, we could see the consequences of the previous day's spillage on the fauna. Whilst a few, predominantly small, sea anemones remained healthy, most of the larger ones were wilting or, like the sea urchins with which they had shared space on the surface of the wooden piles, were lying dead on the surrounding sand. Yet despite the poison still being much in evidence, starfish had arrived in hundreds, to join those already present, carpeting the sea bottom and feasting on those creatures seemingly less tolerant than themselves. Swimming crabs and squat lobsters too had made the trek from the surrounding area to join them, in what seemed to us to remain decidedly unhealthy waters.

In short, a few minutes of observation provided a window on the physiological diversity which animals exhibit. The deaths of many of the large sea anemones and all the sea urchins demonstrated that physiologies which had proved sufficient to enable their survival to adulthood could none the less be overwhelmed. The persistence of small anemones suggested that the timing of extreme environmental events might be crucial in enabling survival, and the persistence of some large anemones suggested the existence of between-individual differences in physiological tolerances or capacities, or small-scale variation in individual circumstances. Finally, the deaths of sea urchins, the survival of only some sea anemones, and the invasion of starfish, crabs and squat lobsters suggested the existence of substantial variation between species in physiological tolerances or capacities.

Although our research interests have at one level been largely unrelated (one of us is a physiologist and the other an ecologist), we have both found the diversity of capabilities of animals in interacting with their environments a source of endless fascination. Whether one considers this diversity at the scale of newborn crustaceans living in the brood pouch of their mother, or at that of latitudinal variation in avian species numbers, many of the fundamental questions remain the same. What level of physiological variation is there? How does it change in time or space? Why does it exist? What are its implications?

This book is a result of this shared response, and the recognition that to date there has been no one volume which provides an explicit overview of the physiological variation which animals display, the patterns in that variation, and the mechanisms and ecological implications of those patterns. It is our attempt to begin to redress that situation. The enormity of such a task means that it is indeed only a beginning. In search of a broader perspective, we have very probably drawn at least some conclusions which in time, or with a more detailed knowledge of the particular topics concerned, will prove to be false. This is inevitable given the breadth of our considerations, and we beg the forbearance of those already more familiar with these particulars.

While this is a new book, there is arguably a sense in which it is already an old one. We have been constantly aware that in writing it we have been 'standing on the shoulders of giants'. We have frequently depended on the work of the greats in the history of ecological physiology, as well as innumerable unsung heroes and standard-bearers. Trawling the length and breadth of the ecological physiology literature, as we have had to do, provides a salutary lesson in the quantity of the work which has been done, the high quality of much of it, and the idiosyncrasies that result in some significant contributions lying uncited and forgotten and some becoming established classics in the field. Of course, whilst the work of others provides the foundation for much of what follows, this need not imply their agreement with all of what we have to say. Indeed, we interpret the outcomes of some studies rather differently from the way in which they were interpreted by the original authors.

Not only does the writing of books rest on the work which has gone before. Books are seldom written in isolation from the influences of colleagues and friends, and this one has been no exception. In particular, we are grateful for those who took time out from busy schedules to read and comment on the draft manuscript, namely Steven Chown, Susanne Eriksson, Sian Gaston, Dave Morritt, Alan Taylor and Phil Warren, also Lorraine Maltby for her comments on Chapters 3 and 4. It goes without saying that they did not always agree with everything they encountered and we did not agree with all their observations. We bear the responsibility for having decided who was right! Others having a less direct input, but none the less making important contributions, intellectual or otherwise, include Stuart Anderson, Susanne Baden-Pihl, Tim Blackburn, Warren Burggren, Andy Hill, John Lawton, Brian McMahon, Phil Rainbow, Tobias Wang, Steve Warburton and Roy Weber. We are also grateful to Noeleen and Dave Morritt and Tom and Margaret Lochhead who provided us with a stunning working environment on the Isle of Cumbrae and at Mouswald on the coast of the Solway Firth, respectively, and to the staff of the *Rising Sun* (where variance in the price of two pints is considerably less than that of a pint and an orange juice). Ian Sherman of Blackwell Science has been an admirable source of encouragement and enthusiasm. As always, we value the support provided by our families, particularly as we battled to juggle completion of the book with other commitments.

J.I.S. & K.J.G.
Sheffield, February 1999

Chapter 1: Introduction

'. . . every physiologist knows what he [or she!] is doing. He may even be convinced that he knows why he does what he does. But does he really recognise the broad implications of his research?' [Florey, 1987]

Diversity or variation is a key characteristic of animal life. The best estimates suggest that there are about 10 million extant animal species, and perhaps many more (Hawksworth and Kalin-Arroyo 1995). These embrace a formidable range of variation in evolutionary history, genetics, morphology, physiology and ecology. On average, each of these millions of species comprises, at least, thousands of individuals, spread across hundreds, or many hundreds, of square kilometres of the landscape or seascape, or through the enveloping media. In most cases, these individuals will have a unique genetic make-up, and, if only in the details, a unique morphology, physiology and ecology. They will, through their life span, encounter a set of environmental circumstances (abiotic and biotic) which is different from that of their conspecifics, perhaps only mildly so, perhaps severely. They may respond in different ways. Their life stories will certainly not be precisely replicated.

The study of biology is the study of the diversity of life. Some facets of this diversity have been investigated exhaustively, others more superficially. With regard to physiological diversity, on the one hand the numerous adaptations of animals to their environment have been the focus of the attentions of innumerable investigators. This has yielded remarkable insights into the variety of ways in which different animals cope with their environment, as well as generating a voluminous literature. On the other hand, the patterns in physiological diversity more broadly, the mechanisms generating them, and some of their significant implications, have seldom been explicitly enunciated. This book is a step towards resolving this circumstance. More specifically, it attempts to provide an overview of physiological diversity and its ecological implications.

1.1 Physiological diversity

Physiological diversity (or physiological variation; the terms will be used interchangeably) is the variability in physiological characters (or traits) among animals (or plants and other organisms). Physiological traits are regulata and tools of regulation, performance and tolerance. They include such things as thermal tolerance, rate of aerobic metabolism, osmoregulatory ability, haemoglobin–oxygen affinity and resting membrane potential. Like all other biological variation, physiological diversity is the result of genetic, developmental or environmental influences (or some combination thereof).

This diversity is perhaps best viewed as being structured in a hierarchical fashion, in which each level of organization is sequentially nested within, and contributes to the integration of, higher levels. Such structure is a basic fact of life and a common feature of biological investigations (Feibleman 1955; Rowe 1961; Allen and Starr 1982; Eldredge 1985; May 1989). From the present viewpoint, the most significant of these levels is the individual, population, species and assemblage ('assemblage' is used here to distinguish any between-species comparison, whether the species be close or distant relatives, and regardless of their number).

1.2 Antecedents

In the first edition of what is arguably one of the best, and perhaps most significant, textbooks on comparative animal physiology, Prosser (1950) made no overt reference to physiological variation or diversity. However, the concept permeates the whole of the introductory chapter. Five years later, in a key review paper entitled 'Physiological variation in animals', Prosser (1955) explicitly pointed out that variation in the physiological traits of species and subspecies had, up until that time, seldom been studied systematically. He drew attention to this neglected aspect of physiology, even if he did not completely redress the imbalance present in earlier works (e.g. Barcroft [1934, p. 2]), recognized the extent to which physiological processes vary between animal groups and that there are invariably costs associated with variation, but went no further; Huxley [1942] has a section entitled 'Physiological and reproductive differentiation', but does not really touch on any of the points identified by Prosser). From around that time and for the next 30 years or more, both in primary literature and in textbooks (e.g. Prosser 1950, 1955, 1957a,b, 1958, 1960, 1964a,b, 1965, 1973, 1975, 1986, 1991; Prosser and Brown 1961), Prosser continued to highlight and bring to the attention of physiologists the importance of physiological variation in its own right, particularly with regard to tackling ecological and evolutionary questions, although, as he observed (Prosser 1964a), medical physiologists have long recognized the implications of physiological variation for correct diagnosis (e.g. Harvey 1628, p. 127; Landois 1885, p. 437; Evans 1945, p. 588; Bell *et al.* 1980, p. 409; Case 1985; Malik and Camm 1995). He also emphasized the significance of knowing how such diversity is distributed between various hierarchical levels (Prosser 1964a, 1965).

While the part of Prosser's message that emphasized the importance of taking a hierarchical approach to physiological diversity was not completely ignored, arguably to date very few of the different levels of physiological diversity that exist have been investigated in detail. This lack of attention may be attributed to the fact that most physiologists do not work at more than one (or two) hierarchical levels and/or that they have failed to consider physiological variation as something important in itself. Rather, such variation is often treated as a source of experimental variation to be minimized, or simply as a 'problem' which the regulation functions of an organism have to 'solve'. To describe this as a failure is perhaps too strong, given that many physiologists were not asking questions concerning the variation they encountered. Rather, they were (and still are!) posing the types of questions that demand that experimental variation be minimized if understanding of physiological mechanisms was to progress at all. This said, a number of physiologists who were interested principally in the ecological and evolutionary implications of their studies, perhaps following the tone set by Prosser, have at various times identified the investigation of physiological diversity at different hierarchical levels as a high priority (e.g. Bartholemew 1958, 1987; Vernberg 1962; Mangum 1963; Barnes 1968; Bennett 1987a; Feder 1987a; Aldrich 1989, 1990; Bennett and Huey 1990; Garland and Adolph 1991).

1.3 Links to ecology

When ecology began to emerge as a distinct discipline it was, in some ways, scarcely distinguishable from what would today be recognized as a comparative approach to ecological physiology. Pearse (1939, p. 100) considered 'Physiology [as] the keystone of ecology'. Physiological variation was flagged as important to an understanding of ecological processes and patterns (Liebig 1840; Davenport 1897; Shelford 1911, 1930; Adams 1913; Allee 1923; Pearse 1923, 1939; Allee *et al.* 1949; Hesse *et al.* 1951; Andrewartha and Birch 1954). Ecological physiology around this time considered the whole series of phenomena of heredity, variation and evolution as being primarily physiological in nature, and suggested that they could not be understood until they were approached from that standpoint (e.g. Rogers 1938). Indeed, some of the thinking concerning physiological diversity was cast in the form of 'laws'. Liebig's (1840) 'law of the minimum', for example, formulates the principle that the factor in which an animal can stand least variation is usually that which limits its survival (still employed as a 'useful' concept 100 years later; Pearse 1939, p. 563; Bartholemew 1958).

A past president of the Ecological Society of America, W.P. Taylor (1936) in his inaugural lecture entitled 'What is ecology and what good is it?', had cause to note (complain?) that 'some physiologists have defined ecology as part of general physiology while others have regarded the two as identical'. However, to a large extent animal ecology and ecological physiology parted company, perhaps as the former began to concentrate more on biotic interactions (competition, predation, etc.) as a way of understanding and explaining animal distributions, and the latter focused on determining the basis of the physiological mechanisms involved in attaining constancy in an animal's internal *milieu* (albeit, in the case of ecological physiology, using animals from different types of environment).

One of the principal aims of ecological physiology has been, and continues to be, to elucidate the implications of physiological diversity, or equally the lack of it, for ecology, namely the differential fitness and the distribution and abundance of organisms (Bartholemew 1958; Vernberg 1962; Vernberg 1968; Huey and Stevenson 1979; Conrad 1983; Kingsolver and Watt 1983; Kingsolver 1985, 1989, 1995a,b; Feder *et al.* 1987; Calow 1988; Pugh 1989; Spotila 1989; Huey and Bennett 1990; Feder and Block 1991; Hoffman and Parsons 1991; Huey 1991; Feder 1992; Fields *et al.* 1993; Johnston and Bennett 1996; Bennett 1997; Somero 1997; Wood and McDonald 1997; Mangum and Hochachka 1998). Attaining such an objective requires the documentation and explanation of patterns in physiological diversity not just within one or two hierarchical levels, but at a number of levels. The extent to which this aim has been successfully addressed is a matter of debate. Indeed, while some physiologists would consider that there is a reasonable understanding of physiological diversity and ecological physiology, such optimism is not shared by all ecologists (e.g. Root 1988a,b; Gaston 1990; Lawton 1991; Brown *et al.* 1996; Chown and Gaston 1999).

From the perspective of an ecological physiologist, some very real obstacles undoubtedly exist to establishing an understanding of physiological diversity. At this point, two major issues can be recognized as being particularly striking. First, as mentioned above, physiological variation occurs at multiple hierarchical levels, from individuals to species assemblages. This means that great caution should be exercised in identifying the sources and determinants of such variation. While the extent and pattern of variation may be studied at a particular level in the hierarchy, this does not necessarily mean that they can be satisfactorily explained at this level. For example, physiological variation observed between two or more species could potentially derive from differences in the geographical origins of the individuals being compared, from differences in the numbers of populations considered, or from differences in the developmental and environmental histories of the individuals. Alternatively, they may reflect genuine fundamental between-species differences in physiology.

Second, historically, where physiological variation has been examined, there has been an emphasis on that which occurs at just two hierarchical levels, the population and the species. Understanding of variation at other levels is thus especially poor (cf. Feder 1987a). Moreover, even for those levels at which variation has been studied more broadly, present understanding remains strongly biased. Knowledge of animal ecological physiology, detailed though it is at the species level, is based on examination of comparatively few species (Prosser 1964a; Florey 1987; Feder 1992). Such a pragmatic approach takes cognizance of the facts that experimental material must be plentiful, relatively easy to maintain or manipulate, and should demonstrate a particular extreme phenomenon of interest to the investigator (collectively known as the Krogh principle [Krebs 1975]—for every problem there is an animal on which it can most conveniently be studied). It has yielded major benefits to our understanding of pure and applied physiology. The assumption that there are common patterns underlying animal function has allowed the use of 'typical' animals in studies. But how typical are these species? (see also discussion of this point under the heading 'the myth of the "typical" amphibian' in Feder 1992). And to what extent can we profitably extrapolate data obtained from them in order to construct hypotheses on the physiological bases of ecological processes and patterns? In short, there is possibly a severe sampling problem.

These two obstacles to an understanding of physiological variation mean that there are precious few works that provide an overview of the basic patterns and their relation to one another, to say nothing of understanding of such patterns (see Calow 1988; cf. Lawton's (1996) views on the lack of a complete catalogue of patterns in ecology, and his lament that, 'there is no one place to ... get a sense of the main empirical framework of our subject'). There has been a stream of very good texts in comparative animal physiology and environmental/ecological physiology, each of which recognizes the importance of taking cognizance of ecology (e.g. Scheer 1948; Prosser 1973, 1991; Hill 1976; Hainsworth 1981; Hoar 1983; Withers 1992; Randall *et al.* 1997; Schmidt-Nielsen 1997; and all of their various editions). However, nearly all essentially draw their examples from a heavily biased set of species or between-species comparisons.

The lack of a broad overview is of particular concern at the present time. It could be argued that some of the key questions in ecology concern large-scale patterns in macroecology (what are the mechanisms underpinning geographical scale patterns in assemblage structure?), global environmental change (how will these patterns respond to environmental changes?) and conservation biology (how might assemblages best be managed to ensure persistence in the face of environmental changes?). Ecologists who are interested in these fields often make particular assumptions about the physiologies of the animals (or plants) that they are dealing with (for introductions to these fields see Kareiva *et al.* 1993; Primack 1993; Edwards *et al.* 1994; Gaston 1994; Brown 1995; Heywood 1995; Rosenzweig 1995; Hunter 1996; Moore *et al.* 1996; Gaston and Spicer 1998a). Although the links are rarely explicit, generally speaking these studies acknowledge that physiological variation is important, if not central, to an understanding of macro or geographical patterns in ecology.

1.4 This book

The objectives of this book are three-fold. First, to provide an overview of patterns in physiological diversity, within the framework of the hierarchical structure of this diversity, and of the mechanisms underpinning those patterns. This will be done by drawing on examples of specific physiological traits, and explaining how and why they vary. There are certain traits that comprise the bulk of physiological adaptations in different environments (e.g. respiration, nitrogen excretion, osmotic regulation and responses to temperature extremes; Prosser 1950). As recognized by Prosser (1950), it is these that may provide the basis for the ecological distribution of animals. Inevitably, the literature we draw on will be heavily biased towards our specialities, although we have consciously tried to include examples of as many of these different physiological functions/regulations as are relevant, and encompassing as many different taxonomic groups as possible (ranging as widely as cnidarians and mammals).

The second objective is to explore the ecological implications of patterns in physiological diversity. A non-historical view is adopted, focusing particularly, although not exclusively, on clinal variation with regard to latitude, altitude and depth. Where relevant, applied issues such as the implications of physiological diversity for ecotoxicology, conservation biology and climate change are highlighted.

The third objective of this book is to highlight current weaknesses in the study of physiological variation and to suggest potentially profitable avenues of future research.

While it will be apparent throughout the text that the link between genetic and physiological diversity is ever present, we have as far as possible adopted an organismal approach (the relationship between genetics and environmental stress has been examined by others, e.g. Pantelouris 1967; Johnson 1979; Powers 1987; Hoffmann and Parsons 1991, 1997; Hoffmann and Blows 1993; Parsons 1996a,b, as has the link between genetics and physiology). Although genetic correlations are noted where we have found them, we do not elaborate on these. Neither of us have the genetic expertise to be authoritative on such matters, and for most of the text there is no real requirement to explore the link between physiology and genetics more than we have done so.

On a related matter, some may expect a discussion of physiological diversity to centre on the concepts of 'stress' and a 'stressful environment'. These terms are applied widely in many biological disciplines. While there is often a tacit assumption that 'we all know what we mean', exactly what constitutes stress or a stressful environment is extremely difficult to clarify; indeed, the terms may be pseudocognate, with many users assuming that everyone shares the same intuitive definition. Many of the attempts to elucidate this issue have shown the terms often to be misleading and anthropocentric constructs. This difficulty substantially reduces their usefulness or operational value. Consequently, our focus is on the response of the animals to whatever environment they encounter rather than on the environments themselves (desert, polar, etc.).

The physiological variation to which most attention is given is quantitative. In Chapter 2 we try to determine the principal patterns of physiological variation that exist within an individual, as they express themselves through time, and ask questions about their origin. This chapter concerns issues which are unfamiliar to many ecologists. We would encourage such readers not to be deterred by this; the material provides important background for that more obviously relevant to ecology which is addressed in subsequent chapters.

At any given moment a number of very similar individuals, at the same location, may exhibit a characteristic value for any given physiological trait. However, the value may be different for each one, with the result that there is marked variation within a given population. In Chapter 3 we ask how wide that variation is, and what form the frequency distributions of such variation take. The extent to which changing the environment can and does influence the expression of physiological variation is also considered.

In Chapters 4 and 5 we pose similar questions to those addressed in Chapter 3, only we move up a level in the hierarchy with each chapter, considering first between-population and then between-species differences. In each chapter some attention is given to the need to investigate and assess the ecological importance and/or implications of physiological variation at, and between, the particular level(s) under consideration. The final chapter (Chapter 6) attempts to paint a broad brush picture of physiological diversity and its ecological implications and from that basis, suggests where to go from there.

Chapter 2: Growing, Developing and Ageing

'. . . *the Camel you will be thinking of is an adult, while* Camelus ferus bactrianus *includes the young ones as well as the old senile ones and the embryos. We physiologists are a notorious sloppy lot indeed!'* [Florey, 1987]

2.1 Introduction

From an ecological perspective, the lowest meaningful level of physiological variation is that of the individual organism. That is, in the present context, the individual animal. Such within-individual variation is pervasive. The physiologies of individuals often vary as they develop, as they grow and as they age, minute to minute, hour to hour, day to day and year to year. All individuals grow and develop in one environment or a number of different environments; for few environments are conditions either stable or static, and this instability may itself engender physiological variation. An individual cannot be in two places at once. It can, however, visit both places at different times, and no two places are likely to be absolutely identical.

In this chapter we seek to describe, and where possible account for, key patterns in the development of physiological regulations and functions, and in the physiological variation that is expressed while an individual is growing, developing and ageing. We commence with a brief overview of the importance of ontogenic studies in biology, and for ecological physiology in particular. Next there is a consideration of the genetic and environmental origins of physiological variation, and their interaction, identifying and attempting to characterize the key patterns in within-individual variation as we do so.

2.1.1 Old and new agendas for ontogeny

Much is known about the development of physiological traits in a handful of mammals, principally because of their medical relevance, and birds (mainly chickens), due to their accessibility and economic importance. However, more generally, understanding of the development of physiological regulations and functions is quite literally embryonic, and has not received the attention which it deserves. It is not just that the physiological mechanisms require elucidation, but ecological physiologists are in the embarrassing situation of not even being in possession of basic patterns, as were catalogued for morphological development in the 19th and the beginning of the 20th centuries.

An interest in ontogeny has a long and distinguished history (von Baer 1828; Thompson 1917; Garstang 1922; de Beer 1958; Gould 1977), albeit eclipsed for a time by Haeckel's (1866, 1876) misleading 'biogenic law' of 'ontogeny recapitulates phylogeny'. It is currently a high-profile topic in developmental and evolutionary biology (McKinney and McNamara 1991; Hall 1992; Rollo 1994; McNamara 1995; Ohno 1995; Valentine et al. 1996; Møller and Swaddle 1997; Richardson et al. 1997). None the less, nearly all of the available data are drawn from morphological studies. Little attention is given to the ontogeny of physiological traits, except where this can be derived or inferred from morphological studies. Of two notable sets of exceptions, the first are the works of Needham in the first half of the 20th century. Among other things, these examine the development of metabolism and nitrogen excretion in a wide range of animals (Needham 1928, 1930, 1931, 1938, 1942; Needham et al. 1935; a series of seven papers published in the *Journal of Experimental Biology* between 1925 and 1933 and each entitled 'Energy sources in ontogenesis' are reviewed in Needham's books. See also Abir-Am (1985) for a historical perspective on his work; cf. Greengard (1974)). Second are those of Adolph (1968), who draws together all the relevant disparate literature on the ontogeny of physiological regulations and places it within a framework of his own creation (an agenda for the study of the development of physiological regulations). Adolph's (1968) book has largely been forgotten (although seemingly not by Russian physiologists who, it is claimed, have long had an interest in the relations between ontogeny, evolution and physiology—see Natochin and Chernigovskaya 1997). However, many of his thoughts and conclusions are being unconsciously revisited or reinvented by current workers who are interested in adopting a developmental approach to ecological physiology (Burggren 1992; Spicer and Gaston 1997).

2.1.2 Replicating the individual

Part of the reason that variation at the level of the individual has not received sufficient attention may lie in the fact that while this variation is quite profound, it is extremely difficult to investigate. The major problem here lies in the fact that it is not possible to replicate the individual. This generates a catch-22 situation. In order to determine how physiological traits vary with, or during, development and growth, the traits need to be measured. But the very act of measuring them may generate variation in those same traits at later stages of development and growth (or may, in the case of some techniques, result in the death of the individual, negating any possibility of taking measurements at all at later stages). There are essentially two ways to cope with this difficulty. The first is to use measurement techniques which do not, at least as far as can be ascertained, have such effects. The development of non-invasive methods and chronic implant of measuring devices is increasingly making this realistic for some traits, such as ventilation in embryonic and adult fish (Van Den Berg et al. 1995; Thomason et al. 1996); cardiovascular function in embryonic and adult animals (Burggren 1987; Airriess et al. 1994; Jonker et al. 1994; Burggren and

Fritsche 1995; Akiyama *et al.* 1997; Hokkanen and Demont 1997; Vincent and Leahy 1997); nervous regulation in vertebrates (Cox and Fetcho 1996) and in invertebrates (Turnball and Drewes 1996); excretion in insects (Chen and Schleider 1996) and even determination of glucose in blood (Spanner and Niessner 1996). It will not provide a solution for all other traits.

The second approach is to examine physiological variation between individuals at different stages of development and growth, and to equate the results with the patterns of variation that would be observed for a single individual could these be measured. In practice, heavy reliance is, necessarily, placed on this second solution. Currently, the data are such that in what follows there is little choice but to do the same. However, as shall be seen, much caution is required to ensure that the patterns being observed truly reflect within-individual variation in physiological traits, and not between-individual variation (Chapter 3).

2.2 Origins of within-individual variation

The origins of variation in a given physiological trait can theoretically be partitioned into a number of components. If the environmental and genetic components (and their interaction) could be separated, what would be left is the variation resulting from measurement error and random developmental noise. In the following sections it is implicitly assumed that the last two are relatively minor, or at least consistent, sources of variation compared with genetic and environmental effects. This assumption will be examined in a little more detail when we consider sources of between-individual variation in the next chapter (Sect. 3.5.1).

2.3 Genetically determined patterns in within-individual variation

2.3.1 Anatomical complexity

Burggren (1992) suggests that the reason that most physiological investigations employ mature adults as study material is on the assumption that increasing anatomical complexity leads to increasing physiological complexity. Thus, it is in the adult of a species that the most complexity will be found, and by implication the most well regulated (least variable) physiological processes. This view implies that the embryo, larva or infant is an incomplete organism and so operates in a less fit fashion and less efficiently than the older one—the 'incomplete adult hypothesis'. It is best studied and understood by assuming that it is preparing to be an adult (Adolph 1968, p. 6). However, there is another view which says that 'at every stage the complement of properties and regulations is complete . . . for operation of the body' (Adolph 1968, p. 6). Taking this second position—which can be referred to as the 'physiological competency hypothesis'—the most complex or best developed physiological regulations or functions would not always be expected to be restricted to, or even present in, the adult stages. There is therefore no a priori reason that within-individual physiological variation will decrease as ontogeny proceeds. How much support is there for each of these positions? We will first examine data that lend support to the incomplete adult hypothesis.

Many species develop the ability to regulate key physiological functions during ontogeny (e.g. nervous coordination: Fig 2.1; Ten Cate *et al.* 1951; Peters *et al.* 1960; Jacobson 1991; Bjorklund *et al.* 1992; Preuss *et al.* 1997; Wan *et al.* 1997; respiratory and temperature regulation: see text below and Figs 2.2 and 2.3; osmo- and iono-regulation and water balance: Holliday and Blaxter 1960; Parry 1960; Triplett and Barrymore 1960; Alderdice 1988; Charmantier and Charmantier-Daures 1994). Furthermore, the sensitivity of those regulations to environmental change often decreases as ontogeny proceeds. For example, the sensitivity of metabolism (usually estimated via oxygen uptake) to variations in temperature, oxygen, salinity and humidity decreases with increasing development in many species (e.g. see Rombough 1988a, 1996 (specifically Table 2); Walsh *et al.* 1989, 1991a,b; Thomason *et al.* 1996 for fish examples).

Generally, physiological regulations in the endothermic vertebrates, the birds and the mammals appear and persist (Fig. 2.2a,b) until they begin to disintegrate as a result of old age or disease (Sect. 2.4.5). For instance, with age and growth (increase in body size) an increased effectiveness is evident in the ability to cool the brain in birds (Arad *et al.* 1984), and a decrease in variations in body temperature and metabolic rate in birds and mammals (Eppley 1994). The onset of thermoregulation in birds and mammals, the ability to control the body (usually, but not always, at an elevated temperature) within fairly narrow limits, is correlated with increased powers of heat production (via non-shivering thermogenesis, where brown fat is metabolized to produce heat and shivering), reduction of heat loss (via hair cover, etc.), and the further development of hormonal and neural control mechanisms (Cossins and Bowler 1987). The thermal stability established by this particular regulation is one of the mainstays in the reduction of physiological variation of many other important regulations in individual endotherms. Once regulations are in place, however, physiological integration leaves little room for manoeuvre. Indeed, such integration is a major feature of maturity and before its emergence there is often heavy reliance, both physiologically and/or behaviourally, on parental care. In some ways the embryo or infant is less capable of maintaining and regulating itself than is the adult.

Despite the fact that much of the machinery of physiological regulation is put together during embryonic development, most newborn mammals (with the exception of herd mammals) are generally not able to respond adequately to environmental challenge (Stainier *et al.* 1984; Eden and Hanson 1987; Mortola *et al.* 1989; Schuen *et al.* 1997). This is particularly so in the case of thermal stress, as the newborn at birth experiences probably the most

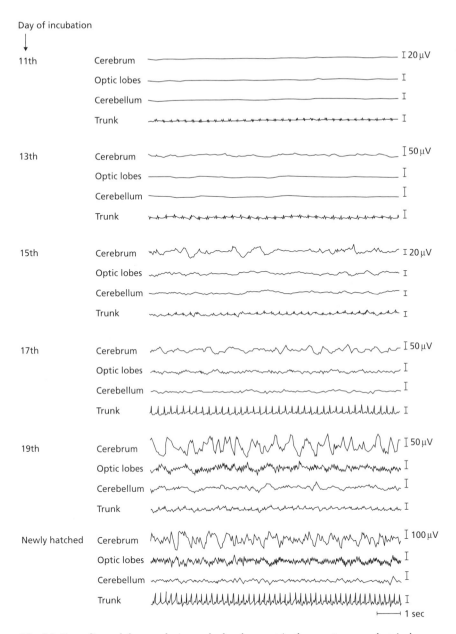

Fig. 2.1 Recordings of changes during early development in the spontaneous electrical activity of a chick brain (11th day of incubation to hatching). On the 11th day the brain is electrically silent, with the activity in the trunk related to heart activity. Activity appears in the cerebral lobes on Day 13 and optic lobes and cerebellum on Day 15. From the 15th day onward electrical activity increases steadily in all parts of the brain until upon hatching the cerebral lobes show slow waves of large amplitude, the cerebellum fast waves of very small amplitude and the optic lobes very fast waves of moderate amplitude. (After Peters *et al.* 1960, with permission from The University of Chicago Press.)

marked thermal change in its lifetime. For example, as with all other mammals, the deep body temperature of the newborn pig decreases sharply from that of the mother's uterus before achieving the adult level some days later (Fig. 2.3). Even in marine mammals that are born at a fairly advanced stage of development, variations in temperature perturb respiratory regulation to a greater extent in the newborn than in the adult (e.g. Hansen and Lavigne 1997).

Similarly, the mechanisms that control ventilation of the lungs develop long before birth in those mammals that have been examined (e.g. sheep, Dawes *et al.* 1972; humans, Boddy *et al.* 1974; Boddy and Dawes 1975), and there are episodic and irregular fetal breathing movements (Bryan *et al.* 1977; Belensky *et al.* 1979; Bowes *et al.* 1981). Upon separation from the pla-

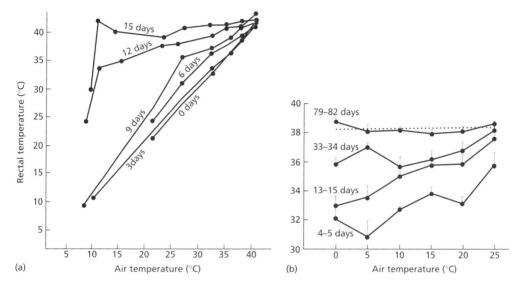

(a)

(b)

Fig. 2.2 Body temperature of (a) wrens *Troglodytes troglodytes* (mean values) and (b) white-tailed jackrabbits *Lepus townsendii* (mean values ± 1 SE) at different air temperatures and different stages of postnatal development. (After Kendeigh 1939 and Rogowitz 1992.)

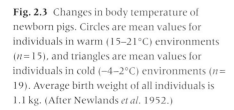

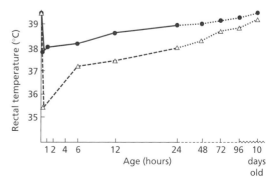

Fig. 2.3 Changes in body temperature of newborn pigs. Circles are mean values for individuals in warm (15–21°C) environments (*n* = 15), and triangles are mean values for individuals in cold (−4–2°C) environments (*n* = 19). Average birth weight of all individuals is 1.1 kg. (After Newlands *et al.* 1952.)

centa, lung ventilation in the newborn becomes continuous, although still irregular compared with the adult pattern (Adamson 1991; Borday *et al.* 1997) (even though the adult control of breathing, via responsiveness to carbon dioxide, seems to be established within 4 h of birth; Praud *et al.* 1997). Greater variation in ventilation in very young stages is also a feature of many lower vertebrates (e.g. the African clawed 'toad' (actually a frog) *Xenopus laevis*; Orlando and Pinder 1995). That in mammals, upon birth, thermal and respiratory regulation, along with many other regulations, are immature should not be surprising given that many of the physiological components of those systems, and related ones, are still developing (Jansen and Chernick 1983; Paton *et al.* 1994; Funk and Feldman 1995; Onimaru *et al.* 1996; Funk *et al.* 1997).

Physiological functions are not just the objects of regulation but are themselves tools of regulation (Adolph 1968). For a particular physiological regulation, say the maintenance of ion balance in mammals, whatever internal processes impact on ion balance are either perturbers or tools of regulation (see Mangum and Towle 1977 who argue convincingly that these two may not be mutually exclusive in some invertebrates). In the case of the tools of physiological processes and regulations, two of the most important players are hormones and enzymes. Decrease in variation of physiological regulations does not necessarily imply that one should also expect a decrease in variation in the tools of that regulation. In fact, arguably, the opposite could be true. For example, the heart of the bullfrog *Rana catesbeiana* becomes more sensitive to the transmitter substance acetylcholine as development proceeds, allowing the individual greater control over cardiovascular activity (Fig. 2.4). On such bases it does not seem unreasonable to link increasing morphological complexity with a decrease in variation in physiological regulations.

These, and many similar, data lend strong support to the incomplete adult hypothesis. However, if this hypothesis has general applicability, the most

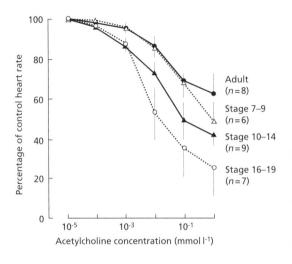

Fig. 2.4 Dose–response curves for the effect of the neurotransmitter acetylcholine on the heart rate of bullfrogs *Rana catesbeiana* at different stages of development. Data are expressed as means ± 1 SE. The number of individuals used is given in parentheses. (After Burggren and Doyle 1986.)

complex or best developed physiological regulations or functions are always expected to be either retained in, or restricted to, the adult stages. This is not invariably so. There is some good evidence that complex regulations of some physiological traits may appear only to disappear again even in early development.

Embryos of the marine isopod *Sphaeroma serratum* only develop the ability to osmoregulate at hatching. Soon after hatching the adult pattern is more or less established (Charmantier and Charmantier-Daures 1994). But contrast this with the situation in the brackishwater amphipod *Gammarus duebeni*. It possesses the adult pattern of osmoregulation, being hyperosmotic in dilute media and isosmotic in more concentrated media, in the earliest embryos examinable (Morritt and Spicer 1995). However, midway through embryonic development, when eye pigmentation is being laid down and the heart is forming, there is a transitory shift to a more complicated osmoregulatory pattern, hyperosmotic in dilute media and hypoosmotic in more concentrated media (Fig. 2.5). It is at this point that the ability to maintain extracellular osmotic control is strongest (i.e. haemolymph osmolality is least variable), in individuals inhabiting an environment which varies with respect to salinity. Such a pattern is frequently associated with attempting to minimize the impact of both low and high salinities in the adult crustaceans in which it occurs (Harris and Aladin 1997), and it has been tentatively suggested that it may be linked with maintaining a more constant internal *milieu* during some of the more critical events of organogenesis (such as heart formation; Morritt and Spicer 1995). An almost identical pattern was demonstrated for the isopod *Cyathura polita* (Kelley and Burbanck 1976).

In the semiterrestrial amphipod *Orchestia gammarellus*, the adult pattern of osmoregulation develops during embryonic development, only to be lost on hatching (Morritt and Spicer 1996a). Consequently, posthatch, the juveniles are retained within the mother's brood pouch, where the osmotic concentration of the water surrounding them is elevated under direct maternal control (Morritt and Spicer 1996b; see also Morritt and Richardson 1998) until the adult pattern (re)develops (Morritt and Spicer 1999). Whatever the reason, or underlying mechanisms, these examples do not square well with the view that an increase in morphological complexity is always accompanied by an increase in physiological complexity, as in each case the most complicated pattern of osmoregulation appears early in embryonic development (with no apparent change in ecological circumstances), only to disappear before, or shortly after, hatching. Likewise, the results of these studies do not lend support to the incomplete adult hypothesis, even if it is sometimes difficult to see how they square with the physiological competency hypothesis.

Such patterns, where complicated regulations appear only to disappear again, are not restricted to invertebrates. The lumpsucker *Cyclopterus lumpus* is a shallow-water marine species which lays its eggs in fairly large rounded masses. Initially, the rate of oxygen uptake of individual eggs is very low and remains so until hatching, even though it increases by a factor of 20 from the

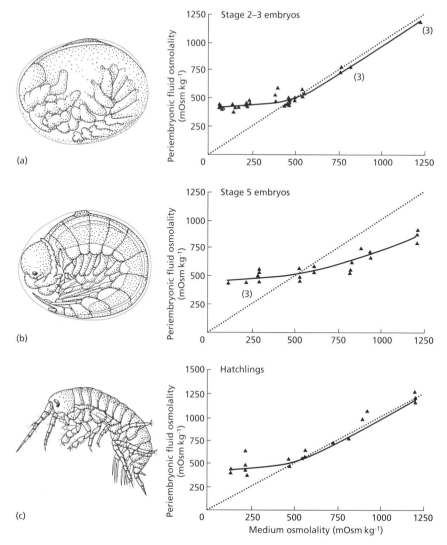

Fig. 2.5 Osmoregulation of periembryonic fluid by (a) Stage 2–3, (b) Stage 5 embryos and (c) hatchlings of the brackishwater amphipod *Gammarus duebeni*. Each symbol represents a single individual ($n=36$ in each case) with figures in parentheses indicating overlapping data points. The broken line represents the absence of regulation. (After Morritt and Spicer 1995.)

moment of fertilization. Upon hatching, oxygen uptake increases dramatically as the active newly hatched larva continues to develop and feed (Fig. 2.6). Respiratory regulation (i.e. the ability of individuals to maintain respiratory independence when exposed to hypoxia, namely low oxygen) is in place very soon after fertilization (6 days or less). Thereafter, as development proceeds, the pattern of regulation becomes weaker (i.e. the P_{crit}, or critical oxygen tension, increases) and at the same time tolerance to low oxygen decreases (Fig. 2.7). In

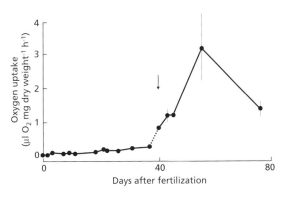

Fig. 2.6 Changes in oxygen uptake of eggs and larvae of the lumpsucker *Cyclopterus lumpus* during development. Values are means (*n* = 3) of group measurements made on 20–30 eggs at the earliest stage of development and 1–3 larvae in the 'oldest' developmental stages. The arrow indicates hatching. (After Davenport 1983, with permission from Cambridge University Press.)

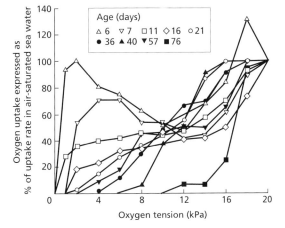

Fig. 2.7 Effect of oxygen tension on oxygen uptake of embryonic lumpsucker *Cyclopterus lumpus* at different stages of development. Values are for group measurements made on 20–30 eggs at the earliest stage of development and 1–3 larvae in the 'oldest' developmental stages. (After Davenport 1983, with permission from Cambridge University Press.)

common with some other fish larvae, by the time the eggs hatch, individuals do not possess any great degree of respiratory regulation, although for some species this regulation will reappear in the adult stages. This means that there is an increase in physiological variation, with respect to respiratory regulation, in developing individuals inhabiting an environment which is variable with respect to oxygen. A similar decrease in P_{crit} with development is found in some other species of bony fish (Rombough 1988a), as well as in embryonic cartilaginous fish such as the dogfish *Scyliorhinus canicula* (Diez and Davenport 1987) (although the presence or absence of respiratory regulation in adult *S. canicula* also seems to be temperature dependent—Butler and Taylor 1975; Thomason *et al.* 1996). Rombough (1988b) shows that the increase in P_{crit} of embryonic steelhead trout *Salmo gairdneri* is related to the accompanying increase in oxygen uptake with development. Initially, the increase is particularly steep until the embryo is able to stir the perivitelline fluid surrounding it, and so enhance diffusion. It could be argued, therefore, that changes in the critical oxygen tension are not due to active regulation. If this were the case then these fish studies do not disprove the incomplete adult hypothesis (reviews of the

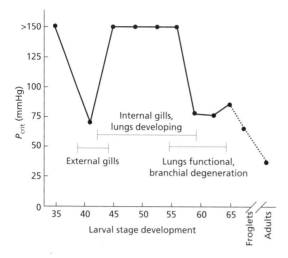

Internal gills, lungs developing

External gills

Lungs functional, branchial degeneration

Larval stage development

Froglets

Adults

Fig. 2.8 Changes in the critical oxygen tension (P_{crit}) of the frog *Xenopus laevis* with development. (After Hastings and Burggren 1995, with permission from the Company of Biologists Ltd.)

effects of hypoxia on fish during early development can be found in Rombough 1988b, 1996). Similarly, in the frog *Xenopus laevis*, the ability to regulate oxygen uptake in the face of acutely declining oxygen tensions (i.e. there is a decrease in P_{crit}) co-occurs with the formation of external gills only to be lost again with their disappearance (Fig. 2.8). The ability to oxyregulate returns just before metamorphosis and thereafter increases with ontogeny (i.e. P_{crit} decreases). In the case of this amphibian, changes in the critical oxygen tension do not seem to be directly related to changes in oxygen uptake.

Even within the birds and mammals there are indications that the regulation expressed by adults is not necessarily the most complicated and/or the most well developed (the two are not always synonymous). Although embryonic birds and mammals are highly dependent on maternal provision, arguably they also possess their own unique physiologies. These are often complicated physiologies, even though they are not able to maintain an existence for the individual without maternal provision; studies of placental function in mammals (Silver *et al.* 1973; Dancis and Schneider 1975; Stave 1978; Mossman 1987) and extra-embryonic structures within bird eggs (Romanoff 1967; Piiper 1978) serve to illustrate this point. But is a particular regulation always more variable at its beginning? In the king quail *Coturnix chinensis* variation in heart rate is greatest at the end of the second week after hatching (Pearson *et al.* 1988). The metabolism of postlarval shrimps of the species *Macrobrachium olfersii*, *M. heterochirus* and *M. rosenbergii* is more sensitive to temperature than at the zoeal (earlier) or adult stages (Nelson *et al.* 1977; Stephenson and Knight 1980; McNamara *et al.* 1985). Finally, while there is a clear trend of decrease in the resting pulse rate of humans with increase in age, arguably the variation is of the same order at each of the ages examined (Fig. 2.9). Clearly, it cannot be said from these studies that there is a decrease in physiological variation with increasing development.

Variation in the expression of the tools of regulation may well closely match shifts in the types of physiological regulation required during development. For example, many tissues have the inherent ability to produce a particular enzyme, or different molecular forms of the same enzyme (isoenzymes or isozymes), and these are expressed at particular times during development. Five isozymes of the enzyme lactate dehydrogenase (LDH_1–LDH_5) are found in various tissues of members of many animal groups. LDH_1 is characteristic of well oxygenated tissues (such as heart tissue) and LDH_5 is characteristic of tissues prone to lack of oxygen (such as skeletal muscle during exercise). The pattern of appearance, prevalence and disappearance of each isozyme during development can vary. In embryonic chickens LDH_1 is predominant, although this changes to LDH_5 immediately before hatching (Cahn *et al.* 1962). Unfertilized and then newly fertilized ova in the mouse, like embryonic chickens, only have LDH_1, but about 9 days after fertilization LDH_5 is predominant (Cornette *et al.* 1967; Rapola and Koskimies 1967). As embryonic development proceeds, there is a general shift back to a predominance of LDH_1, as in the case of kidney tissue (Fig. 2.10), but different tissues follow different itineraries and some

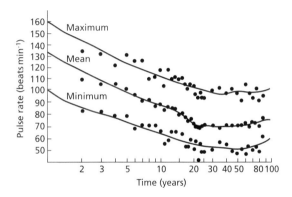

Fig. 2.9 Variation in pulse rate with age in humans. (After Bell *et al.* 1980, with permission from Churchill Livingstone.)

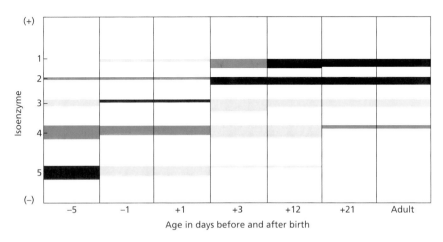

Fig. 2.10 Changing patterns in lactate dehydrogenase (LDH) isoenzymes (1–5) in the mouse kidney during development. The darker the band, the greater the concentration. (After Markert 1963.)

tissues, particularly those that are required to function in periodically or chronically hypoxic conditions, retain LDH_5 in the adult. It is relatively easy in this case to envisage how changing the relative proportions of this enzyme may be important in the regulation of metabolism (aerobic and anaerobic). However, for many of the other changes in isozyme expression that have been recorded during development in a host of different species, while the patterns may be clear, the significance is not (e.g. Moore and Villee 1963). It should be recognized, however, that the identities of the isozymes present at any one time during development are not always pre-set and the onset and quantity of production often have some extrinsic, if not always direct environmental, component. Indeed, sometimes it is only when one knows exactly how the environment affects the expression of different isozymes that sense can be made of the changes in patterns observed during development (see Sect. 2.4.4).

So far good evidence has been presented both for and against the incomplete adult hypothesis. However, for the data that do not support the incomplete adult hypothesis it is not always apparent how much support they provide for the physiological competence hypothesis. Indeed, are the two hypotheses necessarily incompatible? Take the case of the origins of respiratory regulation and subsequent changes associated with the shift from gill to lung ventilation in the bullfrog *Rana catesbeiana* (Torgerson *et al.* 1997). Using isolated brainstems, the activities in cranial nerves (V, VII and X) associated with fictive gill and lung ventilation frequency were measured for tadpoles at different stages of development. Isolates from Stages 2–9 are characterized by rhythmic neural bursts, associated with gill ventilation, which are unaffected by increases in the partial pressure of carbon dioxide (P_{CO_2}) in the surrounding media. By Stages 10–19 isolates are characterized by an oscillating bursting activity, associated both with the gills and the lungs. In Stages 10–14 increasing P_{CO_2} results in a fictive increase in gill ventilation but has no effect on fictive lung ventilation. However, in Stages 15–19 fictive gill ventilation is becoming unresponsive to increases in P_{CO_2} while at the same time such increases are beginning to influence fictive lung ventilation. By Stages 20–25 most of the neural activity measured is associated with fictive lung ventilation and this ventilation is at its most sensitive to increasing P_{CO_2}. On the face of it, these results lend support to the incomplete adult hypothesis, where the bullfrog puts together the adult pattern of lung ventilation during normal development. However, this is to ignore the fact that the larval stage is aquatic and the development of the gills, and their control mechanisms, while being a necessary prerequisite to getting to the adult stage, are also complete in their own right. The larval bullfrog has the correct physiology required to exist as an aquatic larva. This physiological machinery is then lost when the ecological transition from water to air-breathing takes place. In this case, the larval bullfrog is both an incomplete adult and physiologically competent: at the same moment in time it is both an incomplete terrestrial animal and a physiologically competent aquatic one.

Intuitively this must be the case for any individual that undergoes an ecological transition at some time in its life cycle which requires it to survive in a particular environment even while it is still putting itself together—uterus to outside world as in eutherian and metatherian mammals; cleidoic egg to outside world as in prototherian mammals, birds and reptiles; aquatic to terrestrial environments as in amphibians, air-breathing fish and some insects; freshwater to sea water (or vice versa) as in many fish and crustacean species; planktonic to benthic existence as in many marine molluscs, annelids, echinoderms, crustaceans, proto- and urochordates and some fish. That is why comparing the development of individuals to a car assembly line can be so misleading. For many animal groups (although admittedly to different degrees) the car is expected to be in use (often moving through a number of different environments) as it is being constructed. Rather than thinking in terms of the two competing hypotheses outlined above one would perhaps do better to recognize that in terms of physiological regulations, the best developed patterns (least variable) are likely to coincide (i.e. there will be the least mismatch) with the environmental conditions to which they are most relevant at any particular time, and that such relevance will/may change with, or ahead of, major ecological transitions. In evolutionary terms, different selection pressures may be operating on different developmental stages (see Sect. 2.3.2). Many of the studies discussed above that did not lend support to the incomplete adult hypothesis can be interpreted in this way.

2.3.2 Changes in physiological tolerance

Whatever form the relationship between physiological tolerance and physiological regulation takes, there is a long-standing and general perception that the range of physiological tolerance increases during ontogeny (see, for example, Prosser 1964b); juveniles are frequently more sensitive to environmental challenge than adults (e.g. temperature: Vernon 1900; Andrews 1925; Huntsman 1925; Lee and Baust 1987; Stokes and Holland 1996; oxygen: Moore 1942; Eriksson and Baden 1997; Schuen et al. 1997; salinity: Schechter 1943; Anger 1985; humidity: Davis 1974; Morritt 1987; ammonia: Chin and Chen 1988; Young-Lai et al. 1991; pollutants: Rosenthal and Alderdice 1976; von Westernhagen 1988). Closer scrutiny of some of these ontogenic differences indicates that the increase in tolerance is not always progressive. For example, while tolerance to low oxygen by the crab Cancer irroratus changes during early development (there is a three-fold increase in tolerance from newly hatched planktonic Stage I to postmetamorphic benthic Stage V), this increase is not progressive (Vargo and Sastry 1977). Such changes in tolerance as occur can take place over days, such as the development of salinity tolerance by the shrimp Penaeus japonicus (Fig. 2.11), or even more dramatically over a period of a few hours. This is seen in some amphibians which inhabit desert regions of the south-west USA, in which there is an increase in upper thermal tolerance, from 35 to about 40°C, over a period of 24 h, late in embryonic development (Brown 1967).

It would be wrong, however, to give the impression that tolerances always increase during ontogeny. That newborn mammals are remarkably tolerant to hypoxia compared with adults is well established (Reiss 1931; Himwich *et al.* 1942), and it has been known for over 300 years (in 1670 Robert Boyle found that newborn cats survived longer when placed in a vacuum chamber than adults!; Boyle 1670). Indeed, internal hypoxia is thought to trigger a number of key physiological events that surround birth. The importance of enhanced tolerance of low temperatures before efficient temperature regulation is established has also long been recognized (e.g. Edwards 1824, p. 242; Fazekas *et al.* 1941; Mott 1961; Adolph 1969). Furthermore, the knowledge that this tolerance to hypoxia disappears not long after birth is not new (Legallois 1830) and a large number of studies report similar findings for non-mammalian vertebrates (Adolph 1969; Mortola *et al.* 1989). Even in considering changes in tolerance to hypoxia of specific tissues during early development (e.g. nervous tissue; Di Loreto and Balestrino 1997), it is found that there is often a progressive decrease.

Turning to invertebrates, upper thermal tolerance of the starfish *Asterias vulgaris* decreases with increasing size (Smith 1940), as does that of the flatfish *Pseudopleuronectes americanus* (Huntsman and Sparkes 1925) and salinity toler-

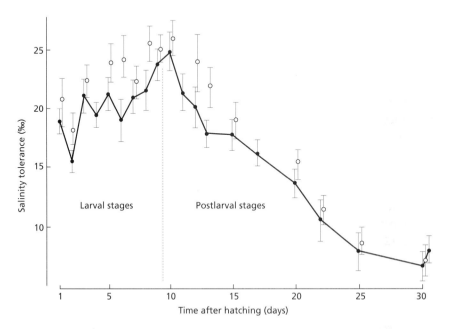

Fig. 2.11 Changes in tolerance to low salinity in the shrimp *Penaeus japonicus* during development. Values are means (95% confidence limits, $n = 10$). The broken line indicates the timing of metamorphosis. LS_{50} values for 24 h (closed circles) and 48 h (open circles). (After Charmantier *et al.* 1988, with permission from the Editor of the *Biological Bulletin*.)

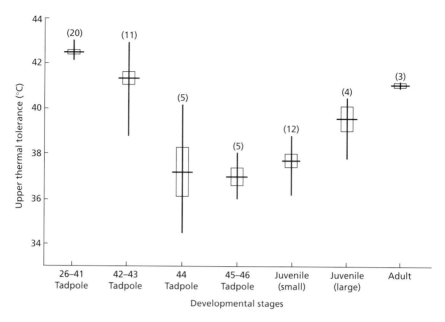

Fig. 2.12 Upper thermal tolerance (CT_{max}) for toads *Bufo woodhousii fowleri* at different developmental stages. Horizontal lines, rectangles and vertical lines represent means, one standard error of the mean and ranges, respectively. Sample sizes are given in parentheses. (After Sherman 1980, with permission from Elsevier Science.)

ance is greatest in larval freshwater shrimps *Macrobrachium petersi* and decreases with increasing development and age (Read 1984). Tolerance of larval and postlarval lobsters *Homarus americanus* to the pesticide pyrethrin (used to treat salmonids for infestation of copepod parasites) decreases with development (Burridge and Haya 1997). Eggs of the Antarctic tick *Ceratixodes* (= *Ixodes*) *uriae*, a parasite of sea birds (also found beneath rocks adjacent to rookeries), are extremely tolerant to low temperatures, but this tolerance decreases with increasing age, with the adult stages being least tolerant (all developmental stages are intolerant of freezing; Lee and Baust 1987). Again, the change is not always progressive. In the toad *Bufo woodhousii fowleri*, although tadpoles possessed a slightly greater upper thermal tolerance (CT_{max}) than adults (42.5°C and 41.1°C, respectively) this decreased to 37°C during metamorphosis and remained depressed throughout juvenile development (Fig. 2.12).

There is an extensive catalogue of changes in tolerance to one environmental factor during ontogeny for representatives of many animal groups. Few investigators have, however, attempted to explore the combined effects of different stresses at any one developmental stage let alone over a developmental sequence. This may not be surprising given that such experiments quickly become unmanageable as additional stresses are factored in. And yet, such studies could be extremely important in revealing not just how tolerances change during development but also by allowing inferences to be drawn about possible environmental interactions and the development of physiological regulations. For example, Costlow *et al.* (1960) constructed 12 different combinations of salinity and temperature for each of which they tested the mortality of five early developmental stages of the estuarine crab *Sesarma cinereum*. Their data are presented in Fig. 2.13, in the form of plots in which percentage mortalities incurred at particular temperature and salinity values are represented by contour lines. Such an approach potentially provides a valuable insight into mechanisms underlying the patterns of mortality observed. The plot for Stage I consists of roughly concentric contour circles, showing that a predicted mortality of 10% or less would be expected roughly in the range of salinity/temperature combinations of 22–36‰ and 22–29°C, respectively. By Stage III there is a quite pronounced distortion of the ellipses, giving way to a ridged pattern in Stages IV and V from which it can inferred that there is a considerable interaction between the effects of salinity and temperature that is evidenced by the absence of a unique maximum or minimum tolerance to either factor. Stage IV, and even more so Stage V, can withstand far greater salinity and temperature extremes than either Stages I or III, although even Stage V seems to be sensitive to low salinities at low temperatures. In sum, although the pattern is complicated there seems to be an increase in physiological competency with development in this species. Similar investigations of combined environmental factors have been carried out for other aquatic animals (e.g. crustaceans: cadmium and salinity — Rosenberg and Costlow 1976; temperature, salinity and naphthalene — Laughlin and Neff 1981; fish: cadmium and salinity — von Westernhagen and Dethlefsen 1975; cadmium, copper and lead — von Westernhagen *et al.* 1979).

Not all patterns of change of tolerance with development are so easy to describe or understand. Although many fish eggs possess relatively narrow tolerance limits, tolerance to cold or heat may in some cases either increase or decrease with, or be independent of development (or size or age; Brett 1970; Elliott 1981; Rombough 1996). Investigations of other animal groups, such as crustaceans and insects, can also point to a similarly confused story (Brett 1970; Precht 1973). Part of the confusion must be due to the use of different techniques and methodologies by different investigators. However, it is unlikely that all of the variation can be explained in this way. That there are inconsistencies should perhaps be expected, and may even be predictable. Physiological traits which have survival value at one stage in an individual's life may not be so important at another (Prosser 1955; Burggren 1992). Intuitively, one might suppose that in ecological terms, greatest significance should be laid on the life stage having the lowest tolerance. However, what is more important is the stage which has the greatest mismatch between the environmental conditions which it is likely to experience and its physiological tolerance *vis à vis* those conditions. For example, adult amphibians and air-breathing fish are often less tolerant to lack of oxygen than early aquatic larval stages (eggs are extremely tolerant), but while the early developmental stages

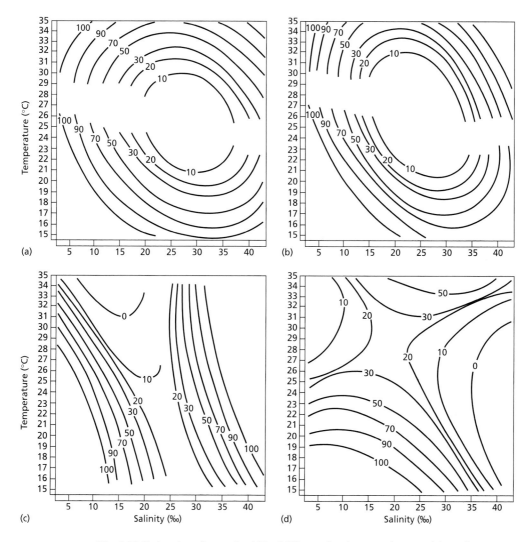

Fig. 2.13 Estimation of mortality (%) of different developmental stages of the crab *Sesarma cinereum*, based on fitting response curves to observed mortality under 12 temperature/salinity combinations. (a) Zoea Stage I, (b) Stage III, (c) Stage IV and (d) Stage V or megalops. (After Costlow *et al.* 1960, with permission from the Editor of the *Biological Bulletin*.)

may actually have to tolerate hypoxia, this particular stress can be escaped in later stages in which the physiological transition from water to air-breathing has occurred (Johansen 1970; Burggren 1984, 1992; Feder and Burggren 1992). Thus, aquatic hypoxia may be a key selection pressure in early developmental stages of air-breathing fish and amphibians, but not necessarily that important in the least hypoxia-resistant adult stage. Similarly, free-living cercariae of the trematode *Himasthla quissetensis* are much more sensitive to anoxia than the endoparasitic adults which are more likely to encounter chronically hypoxic conditions (Vernberg 1963, 1969). Finally, *Drosophila* larvae require a high thermal tolerance if they are to persist at the elevated temperatures that can be found in rotting fruit in the field, but such tolerance is not as necessary for the adult stage which can easily escape from the fruit (Feder *et al.* 1997).

2.3.3 Big individuals writ small?

Is it safe to assume that with respect to a given physiological function or trait immature individuals are merely big individuals writ small? Clearly, size *per se* will have *some* effect on physiological processes even if the answer to this question is broadly affirmative (Peters 1983; Schmidt-Nielsen 1984). As pointed out by Burggren (1992), even without complicatory developmental events (e.g. metamorphosis in arthropods and amphibians), small individuals will differ physiologically from large individuals. For example, the former will have greater weight-specific rates of metabolism than the latter. So the way in which a small individual responds to environmental challenges will be different solely as a result of allometric differences and the physiological demands and constraints such differences impose.

This said, although the physiological implications and consequences of scaling effects (allometry) have received quite detailed attention, nearly all of the work draws on between-species allometry (Sect. 5.3; Peters 1983; Calder 1984; Schmidt-Nielsen 1984). A priori there is little reason to believe that conclusions from between-species comparisons can be applied directly to studies carried out on different developmental stages of the same species. Specifically, one can, for example, ask whether the basal metabolic rate of a 15-g individual mouse *Mus musculus* can be estimated from the allometric scaling of basal metabolic rate across individuals of different species of mammals (even just mice?), or must one possess data on the relationship for *M. musculus* alone?

Heusner (1982) argued that while the mass exponent for metabolism in mammals may have a value close to that predicted by Kleiber (1961) of 0.75, within any particular species the value is likely to be different. Wieser (1984) takes this general conclusion a step further by suggesting that different stages of development may actually require their own allometric equation(s). Regarding our question concerning the mouse, Hou and Burggren (1989) found that for *M. musculus* allometric data derived from between-species studies while they were generally good predictors of cardiac frequency and a number of haematological factors, this was only true for individuals post-weaning. It was only after this stage in development that most of the variation could be accounted for by changes in body mass.

Now take the case of a newly hatched brine shrimp *Artemia franciscana*. It does not have a functional circulatory system as found in the adult because it does not yet have a heart. Cardiac organogenesis, and consequently cardiac function, can only take place with the onset of segmentation, which occurs well after hatching (Spicer 1994). Once the newly differentiating heart appears, a heart beat commences and increases with increasing body size and concomitant differentiation of heart tissue. When cardiac differentiation nears completion and heart growth switches to elongation, there is a marked change in the pattern of heart function; heart rate now decreases with an increase in body mass (Fig. 2.14). Clearly, in this case, early heart function cannot be described purely on the basis of a single power allometric equation.

It has already been noted that many studies of within-individual variation are actually carried out using experimental designs that rely entirely upon between-individual comparisons, and that there may be problems with this. The potential shortcomings of such an approach can be examined by looking in a bit more detail at the work carried out to investigate the development of cardiac function in *A. franciscana* in which one of us (J.I.S.) did just that (lest others think us too quick to criticise). In a subsequent experiment, nine brine shrimp were cultured individually and their heart rates were measured using exactly the same techniques as described in Spicer (1994). Presented in Fig. 2.15 are the data for heart rate (plotted against dry body mass, estimated from the measurement of wet weight at each sampling time) for the two individuals that differ most in cardiac performance. Two things are clear. First, these two extreme data sets fall well within the range of values measured in the previous study. Second, the two data sets show quite different patterns: in one, the

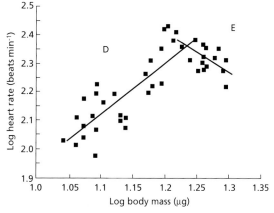

Fig. 2.14 Heart rate as a function of body mass during early development of the brine shrimp *Artemia franciscana* (T = 31°C). Lines fitted by method of least squares for differentiating heart (D) and for heart subsequently growing by elongation (E). (After Spicer 1994.)

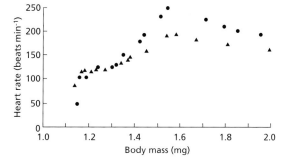

Fig. 2.15 Heart rate as a function of body mass during early development in two individual brine shrimp *Artemia franciscana* (J.I. Spicer, unpublished observation).

initial increase in heart rate is markedly steeper (as is the decrease that follows) than in the other. While the equations presented for the relationships between double logarithmically transformed heart rate and body weight in Spicer (1994) may be used to generate a population response, they cannot be used to predict the (different) responses of these two individuals. The relationship between heart rate and body weight is much tighter for individuals, with possibly some of the variation described in Spicer (1994) resulting at the between-individual level.

Fairly complex changes in heart rate with development have also been recorded for many amphibian, reptile and bird species (Burggren and Pinder 1991; Tazawa *et al.* 1994). So much so, that Burggren (1992) uses changes in heart rate during development in two different anuran species as the basis for the case study that he placed at the centre of a paper entitled, 'The importance of an ontogenic perspective in physiological studies'. The inadequacies of single power functions to describe allometric patterns in changes in physiological traits are amply illustrated by a large number of studies in the literature

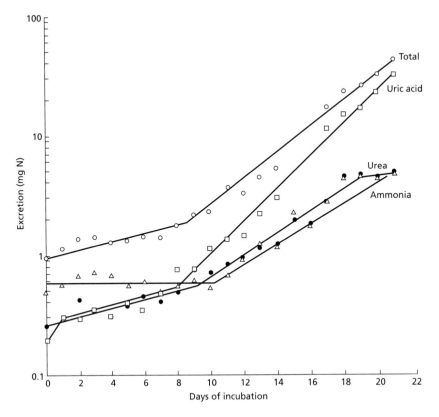

Fig. 2.16 Accumulation of the end products of nitrogen metabolism in embryonic chickens during incubation. (After Clark and Fischer 1957.)

(e.g. metabolism: Chappell and Bachman 1995; digestion: Forbes and Lopez 1989; heart rate: Tazawa *et al.* 1994; Spicer and Morritt 1996).

Nowhere is it clearer, perhaps, that small individuals are not merely big individuals writ small than in the development of nitrogenous excretory function in amphibians, reptiles and birds. In adult reptiles and birds, the primary nitrogenous excretory product is uric acid (Wright 1995). However, in early embryos all three possible products are present; ammonia, urea and uric acid (Needham 1928; Clark and Fischer 1957; Eakin and Fisher 1958). Indeed, from 0 to 6 days of incubation ammonia is the dominant excretory product (admittedly the actual concentrations produced are small), although by Day 20 of incubation nearly all the nitrogen excreted is in the form of uric acid (Fig. 2.16); however, it should be noted that urea and ammonia are still present albeit in considerably smaller quantities. Clearly, early reptile and bird embryos are not even older (bigger) embryos writ small with respect to nitrogen excretion—they are physiologically different beasts.

None the less, some within-individual physiological variation in particular regulations can be predicted on a purely allometric basis, despite the fact that the study individual may undergo quite profound alterations in morphology and possibly even in the physiological tools involved in that particular regulation. Feder (1981, 1982a) investigated rates of oxygen uptake for the larvae of a number of species including the frog *Xenopus laevis,* and assessed the effects of various intrinsic and extrinsic factors on that uptake. He found that while both mass (logarithmically transformed) and developmental stage are correlated with oxygen uptake (Fig. 2.17), examination of the effects of log. mass, stage and the log. mass×stage interaction using stepwise multiple regression shows that mass accounts for nearly all of the variance in oxygen uptake. Even though the size range employed by Feder (1982a) spans a marked metamorphosis in this species, very little of the variation in oxygen uptake could be accounted for by changes in development *per se.* Similar results were obtained for the other species examined. Furthermore, these studies emphasize the importance of using a large sample size, covering as wide a range of masses as possible—doing so increases the precision of the regression and reduces the size of the confidence intervals. Feder's (1982a) experiments also demonstrate that the relationship between oxygen uptake and mass could also be obscured by failing either to allow for, or control, other factors that he shows significantly influence oxygen uptake (i.e. trophic state, time of day, access to air and time-dependent experimental disturbance; Feder 1981) (for similar accounts see also Pruett *et al.* (1991) on amphibian osmotic and ionic regulation, and Arad *et al.* (1984) on brain and body temperature in birds). Even when oxygen uptake (as a measure of aerobic metabolism) can be predicted accurately using single power allometric relationships, as in the previous example, what often cannot be predicted is exactly how, during development, the energy generated is partitioned between major physiological black boxes such as growth, reproduction and maintenance (Fig. 2.18). Such partitioning must have a sizeable genetic component, although it is also clear that partitioning can be open to considerable environmental modification (Townsend and Calow 1981; Sibly and Calow 1987).

Two important points emerge from these considerations. First, some physiological regulations and functions are amenable to prediction using single power functions. However, there is no a priori reason to believe that the form of the patterns detected using a between-species experimental design will hold for a within-individual comparison; indeed, one must be equally careful in using between-individual comparisons to do this. Second, there is no a priori reason to believe that a single power function will be adequate to describe patterns of physiological variation on the timescale of an individual's lifespan.

2.3.4 Must morphological development be accompanied by physiological change?

Many species of anuran amphibians possess external gills for a few days, after which time these are replaced by internal ones (Burggren 1984). For some of those species the larvae will develop lungs, even though they really only seem to contribute to gas exchange around metamorphosis and beyond. Indeed, throughout the life cycle of many amphibians there are fairly complex changes in the relative contributions to the overall functional gas exchange surface made by the skin, the gills (external or internal) and the lungs (Boutilier *et al.*

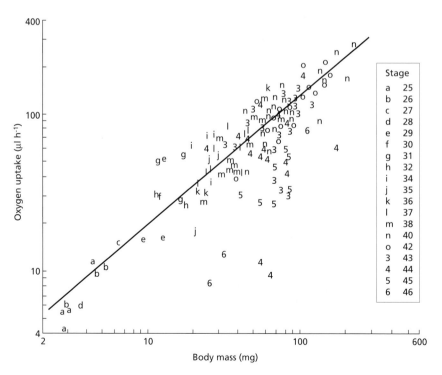

Fig. 2.17 Oxygen uptake as a function of body mass and developmental stage in tadpoles of the frog *Xenopus laevis*. (After Feder 1982a.)

1992; Burggren and Just 1992; Shoemaker 1992). Furthermore, it is also known that profound changes in the oxygen-transporting properties of the blood accompany development in many amphibians, as well as changes in, among other things, osmo- and iono-regulation, acid–base balance, nitrogenous excretion and digestion (Burggren and Just 1992; Hourdry *et al.* 1996). And yet, despite such pronounced structural and functional rearrangement, and the fact that there are often complex changes in mass with development, oxygen uptake for many species can be predicted over a wide range of body sizes and ages on a purely allometric basis (Feder 1981; 1982a; cf. Rombough and Ure 1991; for a comparable situation in fish); while morphology and the tools of physiological regulation may change, the resultant regulation(s) can appear seamless in the face of developmental change.

Nowhere is this more graphically illustrated than in the brine shrimp *Artemia*. Five different sets of structures are associated with osmo- and iono-regulation in an individual brine shrimp. The gills, which are of primary importance in adults, do not appear until well after hatching. The neck organ is a prominent feature that appears, and is functional, just prior to hatching, but disappears before the gills are formed. Although the antennal gland does not become operational until just after hatching, it too disappears while the gills are still forming. The alimentary canal becomes involved in regulation about the same time as the antennal gland, followed quickly by the appearance of a functioning maxillary gland. Indeed, it has been claimed that larval and adult brine shrimp exhibit as much morphological divergence in excretory organs as is found to exist between crustaceans and insects (Conte 1984). And yet, despite such dramatic morphological change, the pronounced degree

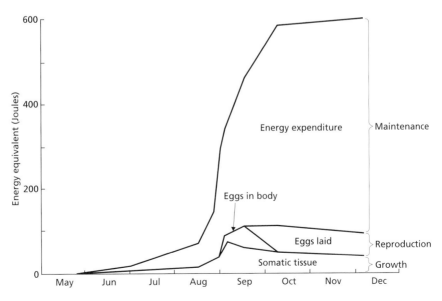

Fig. 2.18 Cumulative energy budget of a female spider *Oligolophus tridens* throughout its life history. (After Phillipson 1981.)

of osmo- and iono-regulation, once established early in development (just before hatching), is indistinguishable from the adult pattern (Conte 1984). Clearly, there must be links between structure and function, but great care must be taken in delineating the exact nature of the relationship. Physiological regulations and functions can proceed apparently undisturbed even in the midst of quite profound morphological and physiological upheaval.

2.4 Environmental modification of the physiological itinerary

2.4.1 Acclimatization and acclimation

Throughout their lives many individuals possess the ability to maintain key physiological functions in the face of environmental change or fluctuation. Although they must be related (exactly how is still not entirely clear), two different phenotypic physiological responses to variation in environmental factors can be distinguished. Modifying the ability to withstand the lethal effects of extreme environmental conditions (singly or in concert) is termed resistance adaptation (Precht *et al.* 1955; Precht 1958). Modifying the rate or performance of physiological functions over a more narrow, and less extreme, range of environmental conditions (again singly or in concert) is termed capacity adaptation (Precht *et al.* 1955; Precht 1958). Capacity adaptations, to varying degrees, tend to offset any disturbance in physiological function accompanying environmental variation (i.e. they include mechanisms of physiological compensation and regulation). The physiological basis of capacity adaptation has received a considerable amount of attention and this material will not be repeated here. Instead, the reader is referred to some key and/or basic physiology texts where the patterns and underpinning mechanisms (particularly with reference to temperature) are presented and discussed in some detail (Prosser 1958; Bligh and Johnson 1973; Precht *et al.* 1973; Cossins and Bowler 1987; Prosser 1991; Hazel 1995; Somero 1997).

In the wild, environmental modification of an individual's physiological processes is termed acclimatization. Although it often involves reversible changes in the physiological phenotype (but not always; Kinne 1962; Rosenberg and Costlow 1979; also Sects 2.4.2, 2.4.3 and 4.2.3), the ability for, and limits of, acclimatization are themselves genetically determined. Acclimatization usually occurs as a result of a complex mixture of changing environmental factors—changes which may be particularly marked in a region with pronounced seasons (Fig. 2.19). Consequently, seasonal acclimatization has received a considerable amount of attention (e.g. Hart 1964; Pohl 1976). Edwards (1824) was one of the first investigators to note the effect of season on an animal's physiology: he observed that the survival time of small birds kept in the same volume of confined air (at 2°C) was better for summer-collected compared with winter-collected individuals. The inference that the metabolic rate of an individual in the winter is greater than that in summer (both measured under identical laboratory conditions) has been borne out by many subsequent studies (e.g. fish—Sumner and Doudoroff 1938; Facey and Grossman 1990; lizards—Tsuji 1988; birds—Weathers and Sullivan 1993; mammals—Bozinovic *et al.* 1990), although this outcome is not invariant, particularly for individuals of species which enter torpor or hibernation in winter (see Sect. 2.5). Seasonal changes in the tools of regulation, such as enzymes (Kent *et al.* 1992; Lin and Somero 1995) and ionic currents (Belkin and Abrams 1998), for example, have also received some attention. Alterations in heat and cold tolerance of vertebrates (e.g. fish—Bulger and Tremaine 1985; Hlobowsky and Wissing 1985; Bulger 1986; birds—O'Connor 1995) and invertebrates (e.g. crustaceans—Layne *et al.* 1987) with season have also been studied in great detail. In general, during one season an individual can tolerate a particular temperature extreme that in other seasons would be fatal. There is even an example of seasonal differences in brain structure and function (Barnea and Nottebohn 1994) which has corresponding effects on behaviours such as singing (Nottebohn 1981) and foraging (Jacobs 1996) in birds.

As well as in response to seasonal changes, acclimatization occurs when an animal invades a new environment. During the breeding season, wild quail *Coturnix coturnix coturnix* move from an altitude of 200m to 1200m to feed, covering a distance of just over 200 km in 6 weeks. This migration is accompanied by marked changes in the oxygen-transporting properties of the blood; there is an increase in haematocrit and in haemoglobin concentration and oxygen affinity (Prats *et al.* 1996). Similar changes are observed in human climbers as they move to higher altitudes (Fig. 2.20). Another example of acclimatization accompanying migration is the change in (selected) enzyme profile during the ontogenic migration of the deep sea fish *Sebastolobus altivelis* (Siebenaller 1984), or the transition from an aquatic to a terrestrial lifestyle (and often back again) resulting in a shift in nitrogen metabolism (from ammonotelism to ureotelism) in some amphibians (Shoemaker 1992).

Each of these examples of acclimatization has been examined over periods of weeks or months. However, it would be wrong to give the impression that such alterations in physiology could only take place over such protracted time scales: in regions where diurnal fluctuations in environmental variables are pronounced there can be acclimatization in the thermal tolerance of some species on a diurnal basis (e.g. fish: Bulger and Tremaine 1985; Bulger 1986).

It would be extremely difficult, however desirable, to replicate such environmental changes in the laboratory. Consequently, physiologists have concentrated on bringing individual animals into the laboratory and varying only one environmental factor while keeping the others as constant as possible. The term acclimation is now employed exclusively for what happens during such laboratory manipulations. Despite the similarities, it is not comparable with acclimatization. For example, the changes in physiology accompanying a

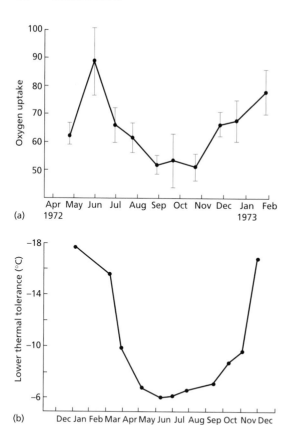

(a)

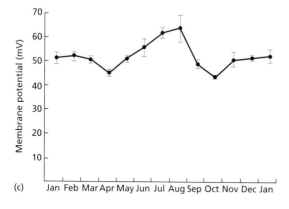

(b)

(c)

Fig. 2.19 Examples of seasonal acclimatization. (a) Routine oxygen uptake by sunfish *Lepomis gibbosus*. Values are means ±2 SE. (After Burns 1975, with permission from The University of Chicago Press.) (b) Cold tolerance of adult barnacles *Semibalanus* (as *Balanus*) *balanoides*. Values are median lethal temperatures (After Crisp and Ritz 1967.) (c) Membrane potential of a recognized nerve cell from suboesophageal ganglia in the snail *Cornu* (= *Helix*) *aspera*. Values are means ±1 SE. (After Kerkut and Meech 1967, with permission from Elsevier Science.)

change in experimental temperature in a laboratory experiment are often quite different from those resulting from seasonal changes, where as well as temperature, other factors such as light and nutrition vary (Markel 1974; Cossins and Bowler 1987; Gatten *et al.* 1988; Seddon and Prosser 1997; although cf. Kleckner and Sidell 1985). Much of the original work carried out by Precht and his collaborators (Precht *et al.* 1955) examines what happens to a number of physiological activities (e.g. oxygen uptake) when organisms are

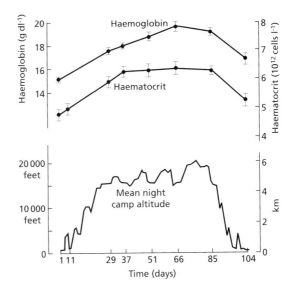

Fig. 2.20 Haemoglobin concentration and haematocrit of the blood of 10 expedition members (aged between 24 and 38 years) during an attempt in 1954 to reach the summit of Makalu, Nepal–Tibet border. Values are means ± 1 SE. Also given is the mean night camp altitude. (After Pace *et al.* 1956, with permission from the American Physiological Society.)

kept at different temperatures in the laboratory (however, patterns of temperature acclimation were in fact studied much earlier; Dutrochet 1837, p. 777; Davenport 1897, p. 253).

There are five different ways in which an individual can respond to changes in a particular environmental factor (Fig. 2.21). After an initial direct effect of the change, the physiological process of interest can return to its original level (part (a) in Fig. 2.21) or at least nearly so (b). However, it can also remain at its new level (c) or in fact continue to deviate from its original level (d). (It is also possible that there could be no response (e).) Presented in Fig. 2.22 are some actual examples of acclimation. As a result of an acute decrease in environmental salinity, from 30 to 15‰, the respiration rate of the blue mussel *Mytilus edulis* initially displays a precipitous decline (Fig. 2.22a). Within a day of the transfer, however, respiratory rate returns to its pretransfer levels. Also accompanying the transfer from high to low salinity is a dilution of the haemolymph, indicated by a decrease in its osmoconcentration. Mussels do not regulate their haemolymph osmoconcentration and so there is no compensation. This decrease is sustained (reduction of salinity by 50% results in 50% reduction in osmoconcentration). Chronic exposure to a higher or lower temperature often (but not always) results in a sustained shift in the upper and/or lower thermal tolerance of an animal (Fig. 2.22b). When a rat is transferred from a high (room) temperature to a low temperature (5°C) in the laboratory there is an initial increase in total heat production. This rate of heat production is not sustained at the same level 60 days after the initial transfer, but it is still significantly elevated above initial and control levels (Fig. 2.22c).

Much of the work on acclimation and acclimatization has been characterized by two implicit assumptions, one potentially confusing and to do with terminology, and the other largely untested. First, although some investigators have used the term acclimation to mean the physiological response to a change in a particular experimental treatment (the sense in which it is used above), others have used it to describe the treatment itself (hence the phrase that recurs again and again, 'we acclimated the animals to') (see the critique in Rome *et al.* 1992, pp. 196–197). Furthermore, there is potential for another misunderstanding to creep in. It is not unusual to describe the 'result of acclimation' on the rate of a physiological process in terms of an animal's ability to compensate. The definition of acclimation, as suggested by the Commission for Thermal Physiology of the International Union of Physiological Sciences (1987, p. 568) is 'a physiological change, occurring within the lifetime of an organism, which reduces the strain caused by experimentally induced stressful changes in particular climatic factors'. This implies that acclimation must always be compensatory. Thus, it could be suggested that in Fig. 2.21, (a) would show perfect compensation, (b) partial compensation, and both would be referred to as acclimation; (c) would show no compensation, and (d) would show reverse compensation with neither fitting the definition of acclimation.

Thus, three features which are quite separate, and should be recognized as such, are confused in much of the literature on acclimation; the experimental treatment, the response of an individual's physiology to that experimental

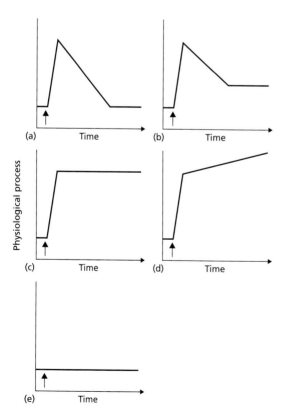

Fig. 2.21 Possible alterations in a physiological process following a change in an environmental factor (indicated by an arrow) such as an increase in temperature or a reduction in salinity. The initial change (in this case an increase, although there is no reason why it cannot be a decrease) is the acute response to the change; what follows is often termed the acclimation response (a–d). There is, of course, another 'response': that there is no physiological change as a result of a change in the environment (e).

treatment, and the direction and benefits of such a response. In the case of the final feature it may not be wise to assume a priori that perfect (or even partial) compensation should be any better or more beneficial to the individual than no compensation. This brings us to our second point.

It has long been assumed that acclimation (whatever the actual response), in at least some cases, is beneficial to the individuals in which it occurs. And yet, despite the great deal of time and energy spent investigating the methods and types of acclimation, until recently the assumption that it has some fitness value has not been subject to rigorous testing (Huey and Berrigan 1996). We

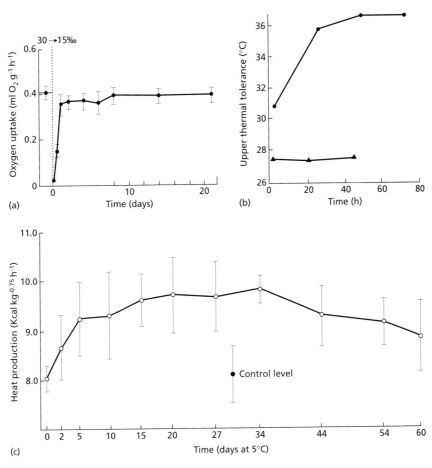

Fig. 2.22 Some examples of acclimation. (a) Effect of an abrupt reduction in salinity on oxygen uptake by the mussel *Mytilus edulis*. Values are means ± 1 SE, $n = 8$ in each case. (After Widdows 1985.) (b) Effect of transfer from 5 to 25°C on the upper temperature tolerance (CT_{max}) of two frog species, *Philoria frosti* (triangles) and *Pseudophryne bibroni* (circles). Values are means, $n = 28$ and 17. (After Brattstrom 1970, with permission from Elsevier Science.) (c) Effect of transfer from room temperature to 5°C on heat production in adult rats. Values are means with ranges ± 1 SD, $n = 40$. (After Cottle and Carlson 1954, with permission from the American Physiological Society.)

will return to this point in the next chapter when examining differential fitness between individuals in a population (Sect. 3.6). What this does mean is that here, and throughout the remaining chapters, when acclimation is referred to, it is to the response of an individual to an environmental challenge, whatever the nature of that response. This limited definition has received some criticism but mainly because it says nothing of the beneficial nature of acclimation (Rome *et al.* 1992, p. 197). Whether or not this criticism is sustainable will largely depend, of course, on whether acclimation is actually beneficial (Sect. 3.6). In conclusion, there is considerable scope for generating physiological variation via acclimation and acclimatization (cf. West-Eberhard 1989), although the relative importance of these as sources of variation has yet to be satisfactorily examined.

From a practical perspective, it has been suggested that by appropriate treatments the occurrence of acclimation could be induced in order to improve the performance of, for example, pest control agents on their release (Huey and Berrigan 1996; Scott *et al.* 1997).

2.4.2 Induction of physiological traits

There are two main ways in which the physiological itinerary of an animal can be modified. The expression of a particular physiological trait can either be switched on or off, or the timing of the onset and/or intensity of that physiological trait can either be delayed or brought forward (Sect. 2.4.4). Such modification can be reversible or irreversible.

Physiological traits can either be switched on or off as a consequence, or as a response to, environmental challenge. For example, in Himalayan rabbits and Siamese cats black pigment is produced only below a certain threshold temperature. Normally only the extremities fall below this, but pale extremities and black on the body have been induced experimentally (Huxley 1942, p. 64). Similarly, in many arthropods when one (or more) leg is removed or lost, there is enough neural plasticity to allow the remaining legs to change walking sequence immediately, allowing locomotion to carry on relatively unimpaired (Prosser 1986, p. 498). Furthermore, regeneration of the lost limb can result in neural regeneration and repair (Cooper 1998). Sometimes switching on and/or off is restricted to a relatively narrow time span, or window, during ontogeny. Such critical windows will be discussed in some detail in the section that follows (Sect. 2.4.3). However, many traits can be switched on or off throughout (most of?) ontogeny. Two particular examples of current interest are the induction of heat shock proteins, which (among other things) appears to be linked with modifications in heat tolerance (see Fig. 2.23; Lindquist 1986; Chen *et al.* 1987; Morimoto *et al.* 1994; Feder *et al.* 1995; Cossins 1998), and the induction of metal-handling proteins (metallothioneins), which have been implicated in tolerance of heavy metal pollution in the environment (Dallinger 1993).

2.4.3 Critical windows

The concept of programming is not new. That is, an early stimulus or environmental challenge, operates at a critical or sensitive period, and results in long-term change in the structure and function of an organism (Lucas 1991). It is, none the less, currently a topic attracting much interest both in medical and ecological physiology (cf. also the consequences of prenatal experiences for later behaviour and neurobiology in Hepper 1987 and Cruikshank and Weinberger 1996, respectively). A controversial book by Barker (1992; see also Barker 1998) claims that factors affecting the human fetus and young at so-called critical or sensitive periods have long-lasting physiological effects, being important causes of later diseases such as heart disease, stroke and chronic bronchitis (diseases normally associated with lifestyle). For example, compromised fetal haemodynamics (blood flow) or nutrition are linked to raised blood pressure in later life.

There is some evidence that a rat exposed to intermittent cold shocks (4°C) in the first week or so of life shows an increase in non-shivering heat production which lasts about 18 weeks (Doi and Kuroshima 1979). Considerably

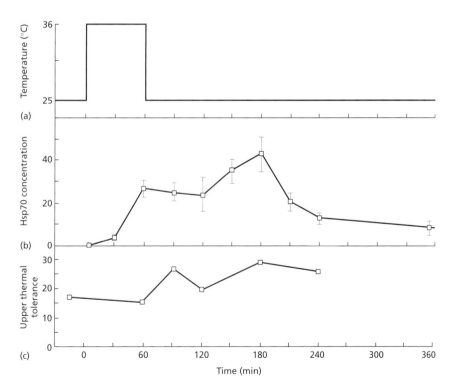

Fig. 2.23 (a) Effect of acute heat shock (1 h at 36°C) and subsequent recovery (at 25°C) on (b) production of heat shock protein (concentration of Hsp70 as percentage of standard) and (c) upper thermal tolerance (LT_{50} (in minutes) at 39°C) in the third-instar larvae of the fruit fly *Drosophila melanogaster*. Values for Hsp production are given as means ± 1 SE. (After Feder *et al.* 1996, with permission from Cambridge University Press.)

more evidence is available for the observation that hypoxia imposed early in postnatal life results in often radical alterations in physiology that can persist throughout the rest of a mammal's life (Strohl and Thomas 1997). Brief hypoxia after birth affects the development of adult ventilatory control, altering breathing patterns (Okubo and Mortola 1988) and causing attenuation of the hypoxic ventilatory response (Okubo and Mortola 1990). Similarly, brief exposure of a newborn rat to hypoxia and auditory stimulation has a lasting impact on ventilatory patterns during sleep in adult rats (Thomas *et al.* 1992, 1995). The normal ventilatory response to hypoxia by an adult rat can be prevented from developing by exposing the individual to high oxygen for the first few months of life (Ling *et al.* 1997). All of the effects described above may be due to alteration in the development of neural structures associated with these aspects of respiration. The first few days after birth seem to be particularly critical for neural development (e.g. Soulier *et al.* 1997) although, as seen above, alterations in the development of physiological function are not restricted to this period. Nowhere is the ability to alter neural regulations early in development more graphically illustrated than in the manipulation of the development of normal sensory inputs in mammals. If the eyelid of a newborn cat is sutured closed in the first 12 weeks of life, thereby depriving that individual of normal sensory input, visual acuity will be lost permanently (Wiesel 1982). The same is not true if an adult is used instead.

Given that physiological itineraries can be altered irreversibly in the laboratory, the question arises as to whether the environment an individual experiences in the wild can act in a similar way. Certainly exposure to hypoxia just after birth, perhaps comparable with at least some of the experimental manipulations employed above, can be a common feature in the life of a newborn mammal, particularly if the birth has been difficult. Turning to the lower vertebrates, there is some very good evidence of irreversible alteration in physiological itineraries. Johnston and his coworkers have for many years investigated the interaction between the environment (in particular temperature) and the developmental itinerary as a way of at least partially explaining some of the physiological (and morphological) variation that exists between stocks of many economically important fish species (Johnston *et al.* 1996). That such within-individual variation could have ecological implications is clear, although it as yet remains untested. For example, Johnston *et al.* (1997) found that the effects of temperature on the relative timing of muscle development in the early embryo of the Atlantic herring *Clupea harengus* persisted into the larval stage (but not necessarily beyond). They note that their study provides a possible physiological mechanism whereby the temperature experienced by the embryo could affect subsequent swimming performance and survival, and hence future juvenile recruitment.

Perhaps one of the best examples of a critical window for physiological development, and one which has clear ecological consequences, is that of sex determination in reptiles as a result of embryonic incubation temperature

(Bull 1980; Ferguson and Joanen 1982; Gutzke and Crews 1988). In fact, the more the possibility of the effect of embryonic environment determining adult physiology and behaviour, and thereby influencing distribution, is examined, the greater the number of examples are found where this is indeed so (e.g. Burger 1991a,b, 1998).

These examples suggest that there is good reason to believe that irreversible alterations in the development of some physiological regulations and functions are more common than has perhaps been generally recognized. If so, this will have clear implications for ecological studies in two ways. First, it may make it much more difficult to distinguish between acclimation/acclimatization and genetic differences in the physiological make-up of an individual (cf. Sects 2.4.1, 3.5). Second, as a matter of urgency, the implications for fitness of such alterations need to be known (see Sects 2.6, 3.6). We will return to both of these points.

2.4.4 It's all in the timing: physiological heterochrony

The timing of the onset of any physiological event can often be altered (which could be referred to as physiological heterochrony) by altering the rate of development *per se*, through changing an animal's environment. For example, culturing the brine shrimp, referred to above (Sect. 2.3.3), at increasingly greater temperatures (at least up to a point) results in an acceleration of development (Spicer 1994). Thus, the onset of cardiac development and function is brought forward in real time as part of the overall developmental itinerary. However, continuing to increase temperature may, as well as accelerating development, have deleterious effects. The rate of development of the beetle *Sitophilus* (as *Calandra*) *oryzae* is greatest at 30°C but mortality is very high. Thus, the range of temperatures over which development can proceed is wide, but the range over which rapid development is possible, without sustaining substantial mortality, is much narrower (Bursell 1974).

To go a step further, one can ask to what extent physiological development is tied to morphological development, or put another way, to what extent is it possible to uncouple the appearance and development of a physiological regulation or function from morphological development and still end up with a viable organism? Intuitively, many mammalian physiologists react negatively to such a suggestion. They point out that the degree of physiological integration is such in most of the higher animals that attempting to uncouple regulations quickly results in an organism which is not viable. Physiological regulata are not like separate pieces of a mosaic that can be added or subtracted at will. But the issue is how much latitude there is in modifying the mosaic. One way of looking at this is to examine how common is the occurrence of canalization, that is, that developmental processes result in the production of a single phenotype, irrespective of environmental variation (Waddington 1959; Scharloo 1991). As we shall see, it is possible, at least to some extent, to modify the timing and expression of physiological itineraries, even in some of the higher animals.

Larval crabs and lobsters possess little ability to regulate their metabolism when exposed to progressive hypoxia (Belman and Childress 1974; Spicer 1995). However, quite a pronounced pattern of regulation is found in the adult stages, which may periodically experience hypoxia in the benthic environment. The development of respiratory regulation can be correlated with structural (Terwilliger and Terwilliger 1982; Wache *et al.* 1988; Durstewitz and Terwilliger 1997) and functional (Terwilliger *et al.* 1986; Spicer 1995; Brown and Terwilliger 1998) changes in the respiratory pigment haemocyanin. There is a shift from a low to a high affinity for oxygen, a high affinity being better able to ensure adequate oxygen uptake and delivery during hypoxic exposure. The interesting issue that arises is whether the developmental itinerary can be experimentally manipulated, to bring forward the adult pattern of respiratory regulation and the expression of the adult high affinity haemocyanin, or whether the timing of these events is genetically fixed. Certainly, modification of haemocyanin function on exposing individuals to hypoxia and salinity has been shown for adult, but not larval crustaceans (Mason *et al.* 1983; deFur *et al.* 1990). Respiratory regulation can be improved in larval Norwegian lobsters *Nephrops norvegicus* by pre-exposing them to sublethal levels of low oxygen, although not quite to the point of being identical to the adult pattern (Spicer 1995; Eriksson and Spicer, in Eriksson 1998). This improvement in *N. norvegicus* is also accompanied by a slight increase in the affinity of the haemocyanin for oxygen.

Temperature-related shifts in the relative timing of key growth and differentiation events may alter the relative sizes and functional attributes of tissues and organs in embryonic crocodiles *Crocodylus johnstoni* (Whitehead *et al.* 1990). Such variation as a result of incubation environment then feeds into longer-term influences on, it has been suggested, growth and survival.

It is well established that the environment into which the newborn mammal is delivered influences the onset of the adult pattern of thermoregulation, a cold environment serving to delay this (Newlands *et al.* 1952). That prevailing environmental temperature can influence the timing of the onset of thermoregulation has also been demonstrated in many, but not all (e.g. Spiers and Baummer 1990), bird species examined. For example, nestling oilbirds *Steatornis caripensis* of different masses (14–570 g, aged from 1 to 83 days old, i.e. the range spanned almost the entire nestling period) were transferred to one of four different environmental temperatures (held stable for 54 min, the maximum time small nestlings were left unattended by adults) after which individual body temperatures were measured (Thomas *et al.* 1993). As nestlings increased in mass they were better able to regulate body temperature (Fig. 2.24). Furthermore, the mass at which any particular chick could effectively regulate its body temperature increased with a decrease in environmental temperature. At environmental temperatures of 30–35°C nestlings of a mass >95 g could thermoregulate, while at lower temperatures (15–19°C) they required to be >250 g.

Some enzymes appear in embryos in response to the presence of substrate. Injecting phenylphosphate into 14–19-day-old chick embryos induces the early production of the enzyme alkaline phosphatase (Kato 1959), and injection of extra glucose brings forward the appearance (and increases the concentration) of the enzyme glucokinase in the liver of fetal rats (Walker 1965).

The development of some regulations and functions is genetically hard-wired or constrained. That is, they are fixed and are not open to direct modification by the environment. The vertebrate gastrointestinal tract undergoes two main transitions as a result of changing functional demands (at birth or

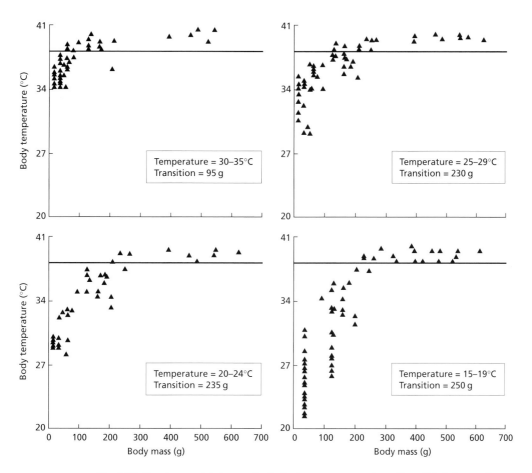

Fig. 2.24 Developmental changes in the body temperature of nestling oilbirds *Steatornis caripensis* kept at four different temperatures. The horizontal line represents the mean adult body temperature. Also given is the body mass at which the 'adult pattern' is achieved. (After Thomas *et al.* 1993, with permission from The University of Chicago Press.)

hatching when it takes over from the placenta/yolk sac, and at weaning when the diet can change) and these transitions are accompanied by the appearance and development of digestive (e.g. Henning 1985) and nutrient transport processes (e.g. Buddington and Diamond 1989, 1992). Testing the theory that sugar (fructose) transporters are induced by exposure to dietary fructose, Toloza and Diamond (1992) showed that keeping rat pups on a fructose-free diet beyond normal weaning did not delay the appearance of transporters. These transporters became functional at their normal time. Testing other dietary items, similar results have been demonstrated for some fish species (Buddington and Diamond 1989), although the onset of glucose transport in the intestine of a lamb prevented from weaning can be delayed if not prevented (Scharrer *et al.* 1979).

While it is apparent that there is some latitude for modifying and even uncoupling physiological itineraries this is certainly not invariant even in the so-called lower animals. In the case of heart function in the brine shrimp *Artemia*, the presence of a functioning heart is tightly constrained by the fact that the heart is formed within thoracic segments. Consequently, cardiac function cannot be brought forward in development in the same way as can changes in respiratory pigment function. The onset of heart function can only be brought forward by early thoracic segment formation, which seems only to respond to environmental factors that influence the whole development (e.g. an increase in temperature results in accelerated growth which brings forward segmentation and hence heart function; Spicer 1994). In the same vein, in the early stages of *Artemia*, exposing individuals to a range of salinities results in an increase in the activity of Na^+/K^+ ATPase in each salinity treatment, but there is no significant relationship between enzyme activity and salinity as is found in the adult stages (Lee and Watts 1994).

Two final examples of genetic hard-wiring involve development in amphibians. There are developmental changes in the activity of a key regulatory enzyme in the ornithine–urea cycle, carbomyl phosphate synthetase I (Wright and Wright 1996). The activity of this enzyme is low in the liver of tadpoles of the bullfrog *Rana catesbeiana* when compared with the metamorphosed adult and is not produced precociously by exposure to environmental ammonia enrichment. Seymour *et al.* (1991) found that the oxygen conductance of the jelly capsule of terrestrial frog *Pseudophryne bibroni* eggs increases with development, but the itinerary is not open to modification by environmental hypoxia or increases in metabolic demand.

In many cases there is a complex interaction between determined ontogenic trajectories for physiological functions and the influence of the environment. Many salmonid fish experience a number of profound morphological and behavioural, as well as physiological, changes in anticipation of seaward migration, but the timing of such changes can sometimes be modified by the prevalent environmental salinity. For example, exposure to high salinities may act as a trigger for the development of salt-secreting mechanisms or anticipatory changes may be reversed by holding fish in freshwater after the migratory period ends (McCormick *et al.* 1985; McCormick and Saunders 1987; Hoar 1988; Primmett *et al.* 1988; McCormick 1994).

In Sect. 2.3.1 we saw that there are often quite complicated patterns of change in isozyme expression with development. However, it would be misleading always to think of the timing of isozyme expression as being completely hard-wired. Although the genetic programme determining the expression of different LDH isozymes nearly always predominates, this programme is also open to environmental influence, as in the case of rats bred at simulated altitude. Mager *et al.* (1967) found that individuals at high altitude possessed more LDH_1 in heart and muscle tissues and the adult level of total enzyme was achieved at an earlier stage in development.

To conclude, many physiological itineraries are hard-wired and are not open to modification by the environment. Their number, however, is perhaps not as great as might be imagined. Heterochrony is considered to be an epigenetic mechanism coupling embryonic development to ecology and vice versa (Gould 1977; Alberch *et al.* 1979; Hall 1992). It is often examined as a between-species phenomenon (although sometimes mixed up with lower level comparisons), using morphological traits. However, as a concept it applies equally well to within-individual variation in physiological traits. In some cases it is possible, at least to some extent, to induce physiological heterochrony within an individual without, apparently, compromising integration to the extent of ending up with a non-viable individual. This means that there is no a priori reason why the development of a mosaic of physiological traits, that together comprise the functioning individual, may not have some sections put in before others. It is not unreasonable to suggest that such flexibility during ontogeny may in some cases, on closer analysis, explain what is presently considered to be between-species heterochrony (Sect. 5.2.4 (ii)). Certainly, such within-individual variability may provide the springboard for evolutionary novelty to arise in some species. The extent to which this is true will remain unclear until more serious attempts are made to understand the relationship between within-individual physiological heterochrony and how that relates to heterochrony at higher levels (i.e. between individuals, between populations and between species; Sect. 5.2.4 (ii)).

2.4.5 Disease and senescence

Everything considered up until this point has focused on physiological variation in a healthy individual. And yet if an individual falls ill, or becomes infected with a parasite, there may well be profound physiological disturbance or alterations. This can be illustrated graphically (Fig. 2.25) using a schematic model originally developed in the context of human health (Hatch 1962). The solid curve defines the relationship between physiology (*x* axis) and pathology (*y* axis). An individual can occupy any point on the curve and can occupy different points at different times. The physiology of the healthy individual, the focus of most of this chapter, is delineated by any point on section A of the curve (normal homeostatic mechanisms, where present, operate, and the individual is not stressed) or if it is stressed for any reason the point may move up into section B (compensatory mechanisms operating). However, once an individual moves out of section A or B, into section C, physiological compensation is no longer possible and the individual is diseased, although over this particular range the disease is curable. For example, fever (linked with increase in body temperature) accompanying infection is a common feature of mammals and birds, and even ectothermic vertebrates such as reptiles (Bernheim and Kluger 1976). Perhaps more surprising, inoculating the migratory grasshopper *Melanoplus sanguinipes* with the microsporidian protozoan *Nosema acridophagus* generates a fever which results in an increase in an individual's preferred temperature by about 6°C (Boorstein and Ewald 1987). This in turn results, under some conditions, in increased survival and growth in infected individuals. In a similar vein, the metabolic rate of individual snails *Lymnaea stagnalis appressa* is depressed, and more variable, as a result of infection by the trematode *Cotylurus flabelliformis* (Duerr 1967).

If, for whatever reason, the degree of physiological impairment increases concomitantly with pathological damage, or a healthy individual contracts a lethal disease, a point which previously had been in section A, B or C could move into section D. Here, there is the breakdown of physiological function and regulations (e.g. temperature and respiratory regulations; Flückiger and Verzár 1955; Grad and Kral 1957), and the pathological damage is irreparable; the individual dies.

If an individual survives disease, competition, predation and/or extreme environmental conditions and accidents, after sexual maturity comes senes-

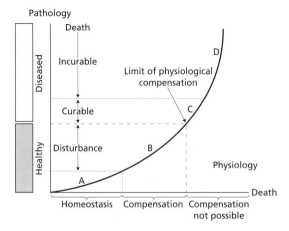

Fig. 2.25 Theoretical relationship between disease and dysfunction. (A) Individual function undisturbed; (B) function disturbed but physiological compensation possible; (C) function disturbed with physiological compensation not possible until pathological damage sustained is repaired; and (D) function disturbed to such an extent that physiological compensation is not possible and the pathological damage incurred is incurable (in practice, such abrupt transitions between states may not occur). (After Hatch 1962, with permission from the American Industrial Hygiene Association.)

cence (old age). This is the period in an individual's life cycle where many of the key physiological functions and regulations begin to fail (often they become more variable), as do the mechanisms preventing and repairing damage (Kirkwood 1981; Finch 1990; Arking 1991; Young 1997). For example, an aged amphibian is considerably less able to generate a lost limb than a young individual (Scadding 1977); an aged mammal cannot regulate its body temperature as well as, and shows less efficient wound healing than, a younger mammal (Goodson and Hunt 1979; Collins *et al.* 1981). Finally, death ensues. This is also equivalent to moving from sections A, B or C to section D in Fig. 2.25. Often it is difficult to pinpoint the exact regulation that failed ('Only rarely can the coroner decide which regulation it was that failed first'; Adolph 1968, p. 131.)

Although senescence is included here under environmental effects, there is also evidence for there being a sizeable genetic component (Finch 1990).

2.5 The importance of behaviour

Individual animals, throughout their life, are involved in a complex suite of behaviours that act towards promoting survival, growth and reproduction (see Fig. 2.26). Such behavioural features affect physiology, directly or indirectly, and there are alterations/disturbances in physiological regulations and functions that accompany them. It may be objected by some that in fact it is the physiology that drives the behaviour. This is particularly so in the case of torpor, dormancy and hibernation, all of which can be regarded as primarily physiological events with behavioural consequences. Whichever way round it is, behaving in particular ways is associated with variations in an animal's physiology. In some cases they ameliorate physiological variation while in others they are sources of variation.

Bennett (1987b), in a paper entitled 'Accomplishments of ecological physiology', overtly recognizes that behaviour is an important component of physiological adjustment to the environment. 'For many animals behavioural discretion is often the better part of regulatory valour' (p. 5). By adopting particular behaviours, for example, moving away from pollutants, becoming torpid or going into hibernation (Churchill and Storey 1989; Kennett and Christian 1994; Clegg *et al.* 1996; also see below) and/or exploiting microenvironments (see Sect. 3.5.2), individuals can mitigate possible disruptive effects of acute (unpredictable), diurnal or seasonal environmental stress (e.g. Fig. 2.27).

Taking cognizance of behaviour is often critical in interpreting the field relevance of physiological studies in the laboratory. For instance, in the laboratory the intertidal prawn *Palaemon elegans* has recourse to an anaerobic capacity that, while not impressive, is more than sufficient for coping with any period of low oxygen they are likely to encounter in the field (Taylor and Spicer 1986). When kept in cages submerged in real intertidal pools they show a similar anaerobic capacity. However, an uncaged individual in a tide pool does not voluntarily stay in hypoxic waters but instead exhibits a partial emersion response (Taylor and Spicer 1988). This response allows the individual to remain aerobic, and negates the generation of an anaerobic energy debt, and the accompanying variation in internal acid–base balance (Taylor and Spicer 1988, 1991). Without the field observation that when a tide pool becomes hypoxic individual prawns exhibit a partial emersion response, one could be drawn to the wrong conclusion concerning physiological variation within those individuals.

As well as ameliorating physiological variation, many behaviours are also the source of such variation. Activity by an individual is usually associated with an increase in rates of aerobic metabolism, and also, in some, recourse to anaerobic metabolism. Other physiological systems and regulations (e.g. acid–base balance, and water balance in terrestrial animals) are also subject to variation with activity. All of these changes have been well documented and will not be repeated here (e.g. Blake 1991; Woakes and Foster 1991; Jones and Lindstedt 1993). Such variability is not of necessity a bad thing. For example, the body temperature of free-living Jackass or African penguins *Spheniscus demersus* varies with (foraging) activity (Wilson and Grémillet 1996). Continued swimming in this well-insulated species results in the body temperature rising above normal resting levels (Fig. 2.28), so that theoretically the individual could be hypometabolic immediately postactivity and remain so until the body temperature drops to a critical level at which point an increase in metabolism will be required to correct it (but see Sect. 3.6 on fitness implications of this feature). To take another example, there are often cycles of body temperature and changes in variation as a result of daily torpor and seasonal hibernation in birds (e.g. Boersma 1986; Calder 1994; Bucher and Chappell 1997) and mammals (e.g. Audet and Fenton 1988; Geiser and Kenagy 1988; Geiser *et al.* 1990; Nestler 1990; Refinetti 1996; Jones *et al.*

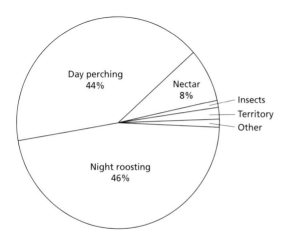

Fig. 2.26 Daily time budgets of a male Anna's hummingbird *Calypte anna*. (After Pearson 1954, with permission from Cooper Ornithological Society.)

1997; Nestler *et al.* 1997; for birds and mammals see review of Geiser and Ruf 1995).

Feeding behaviour (procurement of particular foodstuffs) and the physiology associated with digestion and biosynthesis, result in significant physiological variation. For example, in terms of metabolism, not only is there often some activity-cost associated with procuring food, but in many animals immediately after feeding there is a significant increase in metabolism. This latter feature, admittedly more physiological than behavioural, has been termed specific dynamic action or SDA, and has been studied in individuals of groups as diverse as flatworms (Hyman 1919); crustaceans (e.g. Houlihan *et al.* 1990); fish (e.g. Jobling and Davies 1980; Brown and Cameron 1991) and amphibians (e.g. Wang *et al.* 1995) (for reviews see Jobling 1981); although not all species show SDA (e.g. Amazonian manatee *Trichechus inunguis*; Gallivan and Best 1986). To give a fairly dramatic example of how great, and long-lasting, physiological variation as a result of feeding can be, take the study of Secor and Diamond (1997, 1998) on the Burmese python *Python molurus*. Individuals were fed meals 5–111% of their body mass and rates of oxygen uptake were

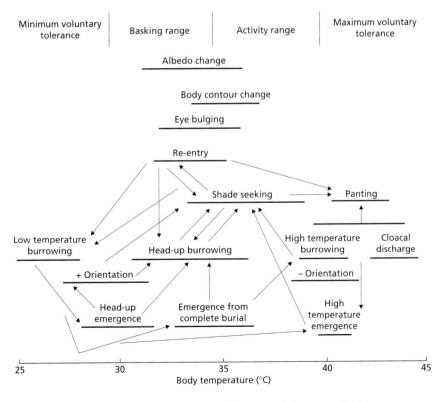

Fig. 2.27 Patterns of temperature-regulating behaviour and the range of body temperature for each pattern, in the lizard *Phrynosoma coronatum*. (After Cloudsley-Thompson 1991, with permission from Springer-Verlag.)

monitored for the following 20 days. There is a peak in oxygen uptake after 2 days (the intensity of which increases with meal size and thus makes a possibly significant contribution to between-individual variation; Chapter 3) which can be 44 times greater than standard rates of oxygen uptake. This is the largest SDA recorded, perhaps with the exception of that measured for race-horses. Furthermore, this large sit-and-wait foraging snake only feeds at long, unpredictable intervals and so between feeding bouts it reduces its standard rate of oxygen uptake by allowing energetically expensive structures (e.g. small intestine) to atrophy. Consequently, the large SDA is attributable in part to the large energy investment required to rebuild these organs on feeding (see similar examples for birds—Janes and Chappell 1995).

There are also a number of other behaviours which could equally well

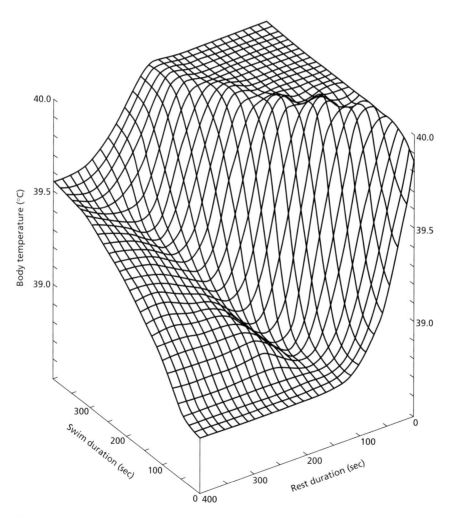

Fig. 2.28 Body temperature of Jackass penguins *Spheniscus demersus* as a function of different swim and rest durations. (After Wilson and Grémillet 1996, with permission from the Company of Biologists Ltd.)

have been presented in the next chapter on between-individual variation, but initially involve responses by an individual. Examples include: aggregation in woodlice and insects affecting water relations (Allee 1926; Willmer 1980); crowding affecting cardiac performance in lizards (John-Alder *et al.* 1996); clustering and huddling behaviour affecting thermoregulation in bats and penguins, respectively (LeMaho *et al.* 1976; Roverud and Chappell 1991); dispersal and reproductive status affecting the metabolism of echinoderms (e.g. Voogt *et al.* 1985), amphibians (e.g. Bucher *et al.* 1982), birds (e.g. Ward 1996) and mammals (e.g. Speakman and McQueenie 1996); foraging (Lighton *et al.* 1987) and copulation (Woods and Stevenson 1996) in insects, agonistic behaviour in crustaceans (Smith and Taylor 1993) and social dominance in fish (e.g. Haller and Wittenberger 1988; Metcalfe *et al.* 1995; Cutts *et al.* 1998), birds (e.g. Roskaft *et al.* 1986) and mammals (e.g. Cane *et al.* 1984) also affecting metabolism. Given this small selection of examples it should at once be apparent that a comprehensive list would effectively be endless.

2.6 The link to fitness

From an ecological perspective it is important to understand whether physiological variation leads to differences in fitness. Discussion of this issue is postponed until Chapter 3, as the fitness of individuals can only sensibly be considered in relation to one another. In principle, physiological variation within the life history of an individual can have profound fitness implications, but considered in isolation it is difficult to demonstrate that this is indeed the case. What happens during development makes the phenotype and determines the quantity and quality of offspring produced. This said, both here, and in many subsequent sections, it should be remembered that the absence of physiological variation could be just as important as its presence. Lack of variation may preclude the appearance of evolutionary novelty and subsequent evolution of particular physiological traits (Maynard Smith *et al.* 1985; Gould 1989; Kirkpatrick and Lofsvold 1992).

2.7 The uniqueness of the individual

'*The slug's a living calendar of days.*' [Thomas, *Here in this spring*, 1952, p. 40]

The expression of physiological variation during development and growth is complex. Genetic and environmental factors, and their interaction, are largely responsible for physiological variation. Developmental changes in physiology, and even in the same physiological regulation (or tool(s) of that regulation) or tolerance, can occur over the time scale of seconds through to decades. Furthermore, interactions with the environment throughout ontogeny may modify an individual's physiology, reversibly or irreversibly. Attempting to describe this variation, let alone account for it, could result merely in a long catalogue of examples from physiologically unique individuals (Fig. 2.29).

However, there are some clear generalities that emerge. It cannot safely be assumed that the adult physiology is representative for any individual developmental stage of a given species. Even in terms of using allometric relationships one is not always secure in the assumption that immature individuals are merely large individuals writ small. In particular, allometric relationships derived from between-species or between-individual comparisons can only be applied with caution to an individual at different developmental stages. While there are many examples of physiological complexity (with a concomitant decrease in physiological variation) increasing with anatomical complexity, this is not invariant. Indeed, morphological variation does not necessarily imply physiological variation. There is evidence that at every developmental stage the complement of properties and regulations is complete for the operation of the body. If the regulation is incomplete and hence variable, this is often compensated for by an increase in the level (and decrease in the variation) of physiological tolerance. Physiological tolerance and regulation often

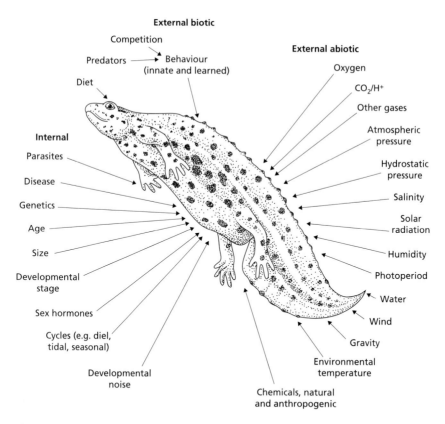

Fig. 2.29 Summary of the types of factors that could potentially result in within-individual variation in a particular physiological trait. The quantity, quality, duration of exposure and rate of change in each of these with time can influence the extent and nature of the variation as can the method of measurement, capacity for acclimation/acclimatization and the interrelationships between controlling and lethal factors.

covary, but the relationship does not always have to be positive. The expression of some physiological traits is hard-wired during development, resulting in very little environmental perturbation, whereas others are extremely labile, and sometimes irreversibly so, in the face of environmental challenge. The behaviour of an individual is a significant factor which can act both to ameliorate and to produce physiological variation. What is certain is that the degree of physiological variation expressed during an individual's lifetime, although there are basic recognizable patterns, is extensive and unique. Consequently, it is self-evident that physiological variation will be encountered at higher hierarchical levels. What is not apparent is the magnitude of this variation.

Chapter 3: Comparing Neighbours

'In some men, there is in such as are brawny and of a harder build, I have found the right auricle so strong that it seemed to be equal in strength of the ventricles of the other men, and truly I was amazed that in different men there should be so great a difference.' [Harvey, 1628, p. 127]

Considerable variation is commonly observed in the values of particular physiological traits between individual animals. For example, in examining data on seasonal differences in evaporative water loss in the aardwolf *Proteles cristatus* (Fig. 3.1) two things are apparent. First, individuals appear to differ in summer and winter in this regard. Second, the between-individual variation is quite marked, particularly in summer, and is sufficient that seasonal differences are blurred. Of course, impressions of the patterns of variation in this case are likely to be influenced markedly by the fact that these data are derived from only five individual aardwolves; this is a respectable number for such a physiological investigation but less than ideal from other points of view.

In another study, this time examining the effect of temperature and exercise on the cardiac output of 13 individual swimming yellowfin tuna *Neothunnus* (= *Thunnus*) *albacares*, Korsmeyer *et al.* (1997) found that heart rate varied considerably between individuals and that there were significant differences in the rate of increase in heart rate which accompanied an increase in swimming speed (Fig. 3.2). The variation was so pronounced that the authors concluded that a function derived from a combined regression analysis of the data for all individuals could not be used as a predictor of heart rate for particular fish, although it could be used to estimate a mean heart rate for a population of yellowfin tuna. Unknown differences in either physical condition or in response to handling stress were suggested as reasons for the variation. However, significant between-individual variation in heart rates has also been found for other fish, both bony (Priede and Tytler 1977) and cartilaginous (Scharold and Gruber 1991). In fact, the magnitude of heart rate variation for yellowfin tuna recorded in the study of Korsmeyer *et al.* (1997) is similar to that found in an earlier study of this same species under different experimental conditions (Bushnell and Brill 1991).

The potential magnitude of between-individual variation in a physiological trait is illustrated graphically in the frequency distributions of heart rate for four individual South Georgian shags *Phalocrocorax georgianus*, continuously recorded over a period of about 5 days (133 023 measurements in total; Bevan *et al.* 1997). All four birds exhibit considerable variation in heart rate; for three

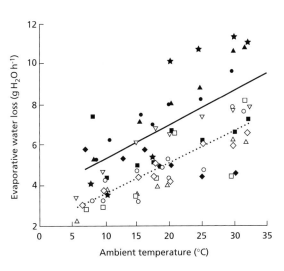

Fig. 3.1 Evaporative water loss as a function of ambient temperature for five aardwolves *Proteles cristatus* during summer (filled symbols, solid line) and winter (open symbols, broken line). Different symbols represent individual aardwolves. (After Anderson *et al.* 1997, with permission from The University of Chicago Press.)

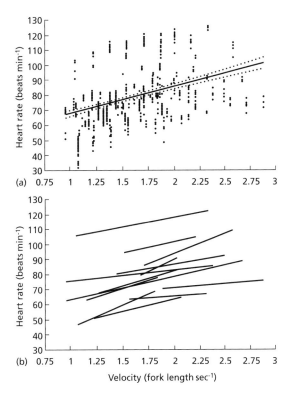

Fig. 3.2 Effect of swimming velocity on heart rate in yellowfin tuna *Thunnus albacares*. (a) Data and combined regression for all 13 fish. Each point is a heart rate recorded at 1-min intervals and multiple points are not indicated. (b) Lines fitted by least mean squares to data presented in (a) for each individual. (After Korsmeyer *et al.* 1997, with permission from the Company of Biologists Ltd.)

of the four birds, heart rate is bimodally distributed, but for one the distribution is trimodal (Fig. 3.3). Generally speaking, high heart rates are measured during the day, when the birds are actively diving, and the low heart rates are recorded at night. Yet again, securing records from such a small number of

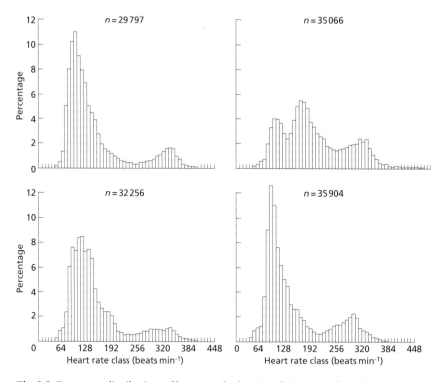

Fig. 3.3 Frequency distributions of heart rate for four South Georgian shags *Phalocrocorax georgianus*. Each heart rate class is expressed as a percentage of the total number of heart rate measurements from that individual (*n*). (After Bevan *et al.* 1997, with permission from the Company of Biologists Ltd.)

individuals may be disappointing but given the logistic difficulties inherent in such a study, not very surprising.

From such studies, and we might equally have chosen from a large number of others, it is clear that between-individual variation is a noticeable feature of examinations of physiological traits. In this chapter, we first examine, for a number of such traits, some of the patterns of between-individual variation which have been documented. We then consider how these patterns alter in the face of environmental change, first in the laboratory and then in the field, and thence the ecological importance and significance (if any) of such physiological variation. Throughout, between-individual variation is considered only within particular populations, and we delay addressing between-population differences until the following chapter. Here, and in subsequent chapters, the term 'population' is used to refer to a group of conspecific individuals co-occurring in a small to moderate-sized geographical area. While a stricter definition is often desirable, as recognized by physiologists and ecologists (e.g. Garland and Adolph 1991; Gaston and McArdle 1993; Haila and Hanski 1993; Wells and Richmond 1995), it is not practical for present purposes.

3.1 Constrained and unconstrained variation

There are two distinct, although intimately related, kinds of between-individual variation. The first is the comparison of individuals at similar stages of growth and development and the second is the comparison of individuals from heterogeneous populations, which may be of different ages and at different developmental stages. These can be termed studies of 'constrained between-individual variation' and of 'unconstrained between-individual variation', respectively (unwieldy as these terms are, they very explicitly distinguish the two sets of circumstances). Of course, this is only a rough-and-ready distinction, as strictly, other than perhaps for clonal organisms, it is impossible to compare individuals identical in growth and development. None the less, it is a useful separation. Almost invariably, restricting comparison to otherwise similar individuals should tend to reduce the between-individual variation observed in physiological traits; at most, the variation can remain unchanged. While constrained between-individual variation is interesting in its own right, and has a number of important ecological implications, from an ecologist's point of view it is arguably unconstrained between-individual variation that is more important. Real populations are typically composed of individuals of differing age and body size.

3.2 Frequency distributions of between-individual variation

The pattern of variation in a physiological trait is perhaps best exhibited by its frequency distribution (remembering that impressions of the shapes of such distributions can be influenced by the way in which they are plotted). This embodies the principal facets of variation, the mean or central tendency, the range, the variance and the skew. In general, it tends tacitly to be assumed that traits exhibit approximately normally distributed patterns of variation, or at least that the distributions are unimodal and approximately symmetrical. There is, however, no a priori reason for such an assumption, beyond the fact that approximately normal distributions are common in nature. Such patterns tend to be produced (i) if there are many single or composite factors influencing a variable; (ii) if the factors are independent in occurrence; (iii) if the factors are independent (additive) in effect; and (iv) if the factors make equal contributions to the variance (Sokal and Rohlf 1981). The extent to which these circumstances pertain to physiological traits is debatable.

In principle, the frequency distributions of physiological traits could exhibit a wide range of shapes. The data may be normally distributed (Fig. 3.4a) or may be significantly skewed (showing asymmetry around a central measure), either to the left (Fig. 3.4b) or to the right (Fig. 3.4c). If unimodal, the distribution peak could be extremely sharp (leptokurtic), or it could have a

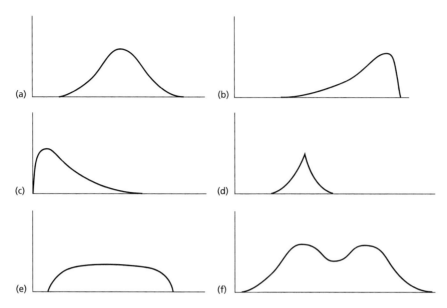

Fig. 3.4 Theoretical frequency distributions. (a–e) are all unimodal. (a) normal distribution; (b) left-skewed distribution; (c) right-skewed distribution; (d) leptokurtic distribution; (e) platykurtic distribution; (f) bimodal.

pronounced plateau (platykurtic) (Fig. 3.4d,e, respectively). Finally, there may be more than one frequency peak (Fig. 3.4f) or even no pronounced peak at all. These different shapes can be characterized using a variety of statistical techniques, and their departure from particular statistical models (in the absence of physiological ones) can be determined (e.g. Sokal and Rohlf 1981; Snedecor and Cochrane 1989; Zar 1996). Thus, it has not been a lack of appropriate tools which has limited the attention paid to the frequency distributions of physiological traits, rather the questions which necessitate that these tools be employed have seldom explicitly been posed.

Two obvious key questions are whether, and if so how, the basic shapes of the frequency distributions of physiological traits vary, and how these shapes change as a result either of experimental manipulation or of environmental change in the field.

3.2.1 Constrained variation

Several things are revealed by collating a number of available examples of frequency distributions for similar individuals of values of different types of physiological trait, be they measures of regulation, tolerance or performance. First, as already noted, there is often significant and pronounced between-individual variation within a population. Second, in most of these cases a strong central tendency is exhibited. This need not always be the case, but it is

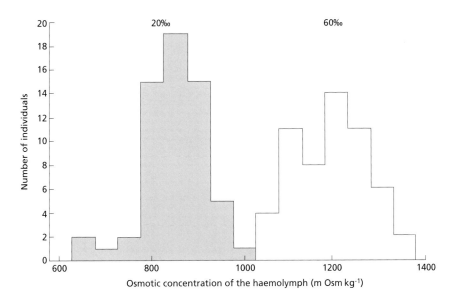

Fig. 3.5 Frequency distributions of haemolymph osmolality for the adult crab *Australoplax tridentata* kept at 20‰ and 60‰ salinity (*n* = 60 and 56, respectively). (After Barnes 1968, with permission from Elsevier Science.)

probably a very general pattern. Third, the actual shapes of the distributions are variable.

The first example depicted here (Fig. 3.5) is, to our knowledge, one of the earliest investigations of between-individual variation in a physiological trait for an invertebrate. It is of the range of freezing point depressions (as a measure of osmotic pressure) for the haemolymph of the estuarine crab *Australoplax tridentata*. Crabs were deliberately collected from within a very small area (<20 m²). Individuals were then carefully selected, standardizing for reproductive and moulting condition, size and sex, and exposed in the laboratory to a salinity of either 20 or 60‰ (2 days at 25°C) (Barnes 1968). The frequency distributions appear approximately symmetrical and normal, although this has not been tested formally. Barnes (1968) concluded that his study highlighted the need for caution in interpreting data obtained from only a few replicates for such a markedly euryhaline crab. He also noted that 'High variability in organisms inhabiting a very variable environment (with respect to salinity) would ensure that under severe, but temporary salinity conditions, some proportion of the total population would survive through to the onset of more favourable conditions; while under a selective pressure in favour of the individuals capable of the greatest osmoregulatory abilities in extreme salinities, further penetration up a salinity gradient may be achieved by this species.'

Another example of a physiological trait with a distribution which is not significantly different from a normal (although this is not necessarily evident on visual inspection), only this time drawing on a much larger sample size, is

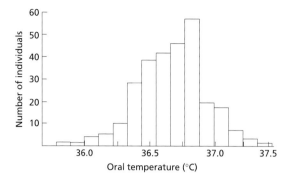

Fig. 3.6 Frequency distribution of oral temperature for medical students ($n=276$) seated in a warm classroom, between 08.00 and 09.00 h. (After Bell *et al.* 1980, with permission from Churchill Livingstone.)

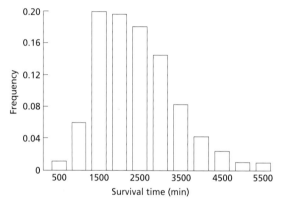

Fig. 3.7 Frequency distribution of survival times of larval sea urchins *Psammechinus miliaris* exposed to 0.1 mg l^{-1} copper ($n=506$). (J.I. Spicer, unpublished observation.)

that of the oral temperatures of 276 medical students (Fig. 3.6). Temperature is strongly regulated in endotherms and yet, despite standardizing on time of day (08.00–09.00 h), sampling location (a warm classroom), and age of the subjects, even here there is still marked variation with the range extending from 35.8°C to 37.3°C, a difference of 1.5°C (see also Fig. 6.1).

Many physiological traits are probably, as with these examples, at least very approximately normally distributed, although, as was the case for haemolymph osmolality discussed above, they often show different degrees of variation from the mean. However, equally cases can be cited of physiological traits where the frequency distribution is plainly not of this form. Data on the survival of larval sea urchins *Psammechinus miliaris* exposed to waterborne copper are markedly right-skewed (Fig. 3.7). Similarly, the frequency distribution depicting tolerance of the aquatic oligochaete *Limnodrilus hoffmeisteri* (collected from uncontaminated water) to a cocktail of metal pollutants is also right-skewed (Fig. 4.3), as are the frequency distributions of the concentrations of trace metals in a number of different marine species (Lobel *et al.* 1992).

Negatively or left-skewed distributions are exhibited by some physiological traits. This is true, for example, for the anoxia tolerance of 143 individual

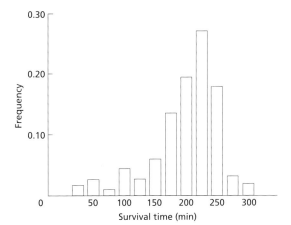

Fig. 3.8 Frequency distribution of survival time in anoxia for intertidal prawns *Palaemon elegans* (n = 143). (J.I. Spicer, unpublished observation.)

intertidal prawns, *Palaemon elegans* (of equivalent body size) (Fig. 3.8). In the case of the prawns, the most tolerant individual survived six times longer than the most sensitive one. The functional basis of these differences may be related to the fact that survival in anoxia (and hypoxia) is positively correlated with initial glycogen content (Taylor and Spicer 1986). Other examples of left-skewed distributions are haemocyanin concentration in shore crab haemolymph (Fig. 3.15a), and heart rate in waterfleas (Fig. 3.30b).

As well as distributions of physiological traits which are approximately normal, right-skewed or left-skewed, there are others which do not appear as easy to categorize (e.g. Fig. 4.6). Some, such as the supercooling point in insects, show multimodal distributions (see Figs 3.12 and 3.14) and although they are often difficult to deal with (Sømme and Conradi-Larsen 1977; Block and Sømme 1982), some investigators have put forward ideas on appropriate statistical analysis (e.g. Aldrich 1987). In many cases, however, the sample sizes on which distributions are based are too small to reveal the nature of the underlying distribution from which individuals have been obtained. Indeed, when measurements have only been made for a very few individuals it is almost impossible to say anything precise about the distributions from which they have been drawn, and there is little merit in attempting to determine their departure, or otherwise, from formal statistical or other theoretical models.

More generally, sample sizes may have a profound effect on perceived patterns of physiological variation. In particular, if a small number of individuals are sampled from a distribution with a long tail, the estimate of the variance is nearly always an underestimate. Occasionally, a huge overestimate is generated, because the distribution of sample variances is itself heavily skewed, which keeps the average estimate unbiased. Let us draw on an example originally discussed by McArdle *et al.* (1990) in the context of determining variation in the size of animal populations, but equally relevant here. If one simulates a population with a negative binomial distribution (right-skewed), with a mean of 50 and a clumping parameter k of 0.1 (which means that it is a quite strongly aggregated distribution), then the true variance is 25 050. Of 1000 samples of size 5 taken from this distribution, 824 estimates of the variance were less than the true value and the median estimate was 1887, less than 8% of the true value. This effect is most marked at small sample sizes; however, when the same distribution was sampled using a sample size of 20, 727 of 1000 samples were still underestimates and the median estimate was 9500, only 38% of the true value. In short, given the small sample sizes typically obtained, most physiological studies stand very little chance of making a reasonable estimate of between-individual variation in a physiological trait, if that trait is distributed according to a markedly skewed function. The vast majority probably substantially underestimate the range of values exhibited by a population.

3.2.2 Unconstrained variation

The impact of variation resulting from age and historical structure on the overall pattern of physiological variation within a population will depend to a large extent on how the physiologies of individual animals change during development (Sect. 2.3) and how such change is influenced or mediated by the environment (Sect. 2.4). For example, if the frequency distribution of normal systolic blood pressure obtained from persons of the same age (18 years, $n = 1216$, i.e. constrained variation) is compared with that obtained from persons encompassing a wide age range (16–40 years, $n = 6000$, i.e. unconstrained variation) it can be seen that there is very little difference between the two (Fig. 3.9). Both show the same strong central tendency, with perhaps some evidence of a slight right-skew, and the range of values is almost identical. In this case, there is no difference between constrained and unconstrained variation, perhaps because blood pressure does not vary with age, at least over the age range examined. The shape of the unconstrained frequency distribution may well be quite different if it were to include fetuses, newborn and older infants, prepubescents and very elderly individuals.

For the woodlouse *Oniscus asellus*, the upper thermal tolerance (CT_{max}) of similar-sized individuals (adults, 20–27 mg) shows a near normal, or at least broadly symmetrical, distribution with a range mainly from 35 to 42°C (Fig. 3.10a). However, it is known that upper thermal tolerance increases with size in this species. Consequently, when examining the upper thermal tolerance (CT_{max}) of a real population, taken straight from the field and consisting of a wide range of different (juvenile and adult) body sizes (median size is quite large, body size is heavily left-skewed), the resultant frequency distribution is more obviously left-skewed (Fig. 3.10b). While most of the individuals examined had a CT_{max} between 35 and 40°C, the complete range extended from 19 to 43°C. Many (although not all) of the very low values of CT_{max} observed, that

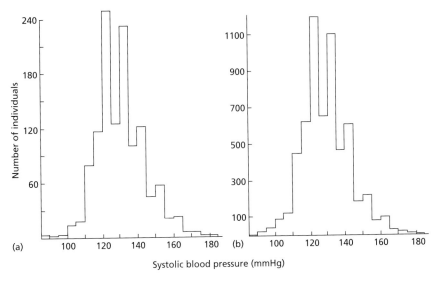

Fig. 3.9 Frequency distributions of normal systolic blood pressure for persons of (a) 18 years of age ($n = 1216$) and (b) 16–40 years of age ($n = 6000$). (After Evans 1945.)

constituted the tail of the distribution, were contributed by some of the smallest members of the population. Unconstrained variation in the upper thermal tolerance of the flatfish *Pseudopleuronectes americanus* shows a similar left skew, with the oldest individuals being the least tolerant (Fig. 3.11). In comparison with the example of human blood pressure discussed above, there is a considerable difference between constrained and unconstrained variation in thermal tolerances for these species.

As mentioned earlier, individuals within a population may differ in many ways which may have important effects on the observed distributions of physiological traits. This is well illustrated by supercooling points of microarthropods, notably mites and springtails, at high latitudes. Bimodal distributions of this trait are regarded as a common feature of field populations of such species (Fig. 3.12; Block and Sømme 1982; Sømme and Block 1982). This bimodality is usually associated with the feeding condition and hydration state of the individuals. Thus, in a population of the collembolan *Tetracanthella wahlgreni*, individuals with gut contents show a range of supercooling temperatures in the range –13 to –4°C (a very few isolated values around –29°C), and with a left-skewed frequency distribution (Fig. 3.12; Sømme and Conradi-Larsen 1977). Individuals without gut contents, however, display a lower central tendency, a greater range of supercooling temperatures (–32 to –5°C), and a heavily right-skewed frequency curve. Gut contents are an important site of nucleation and so promote ice formation at relatively high subzero temperatures. Consequently, it is not surprising that overwintering forms of insects stop feeding, and evacuate the gut long before subzero temperatures are

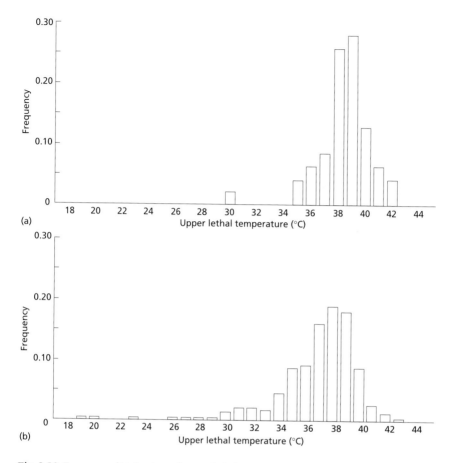

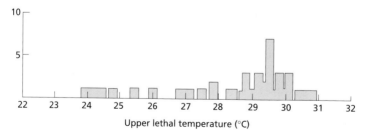

Fig. 3.10 Frequency distributions of upper lethal temperature tolerance of (a) similar sized and (b) all sizes of the woodlouse *Oniscus asellus* ($n = 61$ and 177, respectively). (J.I. Spicer and K.J. Gaston, unpublished observation.)

Fig. 3.11 Frequency distributions of upper lethal temperature tolerance of the fish *Pseudopleuronectes americanus* of different lengths (10–30 cm). Tolerance decreases with an increase in size. (After Huntsman and Sparkes 1925, with permission from the American Physiological Society.)

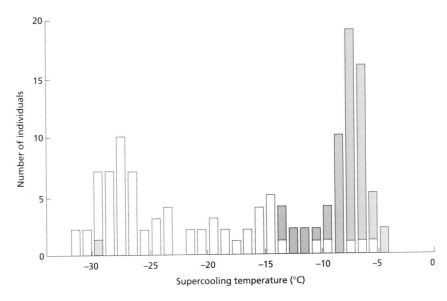

Fig. 3.12 Frequency distribution of supercooling temperatures for individuals of the collembolan *Tetracanthella wahlgreni*, without (open) and with (shaded) gut contents. (After Sømme and Conradi-Larsen 1977, with permission from Munksgaard International Publishers Ltd.)

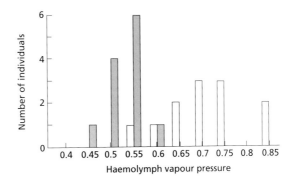

Fig. 3.13 Frequency distributions of haemolymph freezing point depressions from torpid (open) and active (shaded) snails *Helix pomatia* (*n* = 12 in each case). (Data taken from Kamada 1933.)

encountered, thereby preventing inoculative nucleation. This apparently relatively minor difference results in dramatic differences in the patterns of variation surrounding supercooling in this species.

Some good illustrative examples of frequency distributions for unconstrained between-individual physiological variation can also be found in the next chapter, dealing with between-population variation (cf. Figs 4.3, 4.5–4.7). As was seen for constrained variation, data for unconstrained variation within populations exhibit significant between-individual variation, often display a strong central tendency, and may exhibit frequency distributions with different shapes.

Whether constrained or unconstrained between-individual variation within populations is examined or not, there is little evidence for marked discontinuities in the frequency distributions of physiological traits. Rather, variation tends to be continuous. None the less, there have been attempts to categorize individuals within populations into different so-called 'physio-types', distinguished by the state of a physiological trait (Depledge 1990, 1994). For example, marine invertebrates have been divided among physio-types on the basis of variation in the accumulation of trace metals (Depledge and Bjerregaard 1990). In general, the benefits of imposing arbitrary catego-rization of a continuous variable are limited, although it may be useful where there is substantial uncertainty in the measurement of a given trait but errors in placement in categories are likely to be sufficiently small so as not markedly to alter subsequent inferences.

3.3 Experimentally altering between-individual variation

In the laboratory, it is possible to subject a population to a carefully controlled treatment, and then to examine what effect this has on between-individual physiological variation, particularly as summarized by following changes in the frequency distribution of a trait. Besides the obvious null outcome of no change, two distinctly different kinds of results may be observed.

First, the experimental treatment may change the shape of the frequency distribution by changing, directly or indirectly, the physiologies of some or all individuals. For instance, what do we see when we compare the vapour pres-sure of haemolymph from the edible snail *Helix pomatia* while individuals were in torpor (hibernating) with the same individuals a week later when they were active, following the simple expedient of increasing the temperature (Kamada 1933)? The vapour pressure is on average higher for torpid than for active individuals, and the frequency distribution of vapour pressures is flatter in the former case than in the latter (Fig. 3.13; although sample sizes are very small). In the case of the supercooling temperature of the polar insect *Cryptopy-gus antarcticus* (Fig. 3.14), while field-collected and laboratory-kept individuals who were fed moss turf show a left-skewed frequency distribution (or perhaps a bimodal distribution with the right peak far greater than the left one) after 3 days without food, the frequency distribution is right-skewed (or if bimodal the left peak is now the more pronounced).

The second effect of an experimental treatment may be that the shape of the between-individual distribution of a physiological trait could change as a result of selection, that is, because of the differential mortality of some indi-viduals. In reality, even in the laboratory, the imposition of a stress that results in the mortality of some individuals in a population will tend also to result in changes in the physiology of those individuals which persist. A good example arises from recent work on the respiratory pigment haemocyanin. This

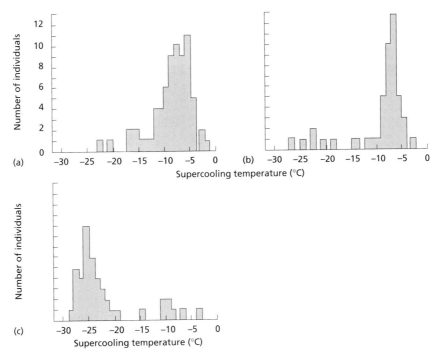

Fig. 3.14 Frequency distributions of supercooling temperatures for *Cryptopygus antarcticus*.
(a) Field-collected individuals ($n=70$); (b) individuals fed moss turf for 5 days after
collection ($n=47$); and (c) individuals starved for 5 days after collection ($n=43$). (After
Sømme and Block 1982, with permission from Munksgaard International Publishers Ltd.)

pigment occurs sporadically within the Mollusca and the Arthropoda, and in
most species the concentration is thought to be extremely variable (Mangum
1997). Looking at a large number of individuals of the shore crab *Carcinus
maenas*, the range of variation is substantial (approx. 10–60 mg ml^{-1} haemocyanin), and the frequency distribution of the concentration is approximately
left-skewed ('before' in Fig. 3.15a). When these same individuals are sub-
jected to severe environmental stress in the form of exposure to hypoxia (40%
of normal saturation for 8 days), two things happen. First, there is significant
mortality, of the order of 30%. Second, as is seen in Fig. 3.15a, among the survivors the haemocyanin concentration is considerably less variable (by a factor
of 2–3) than was the case for all the individuals pre-exposure, the mean concentration increases, and the concentrations are more symmetrically distributed. The increase in haemocyanin concentration with hypoxia is not
unexpected, as it allows individuals to compensate for the reduction of oxygen
in their environment (see also Hagerman and Uglow 1985; Hagerman 1986;
deFur *et al.* 1990). These changes could not occur simply by differential survival of individuals with high haemocyanin concentrations, and no physiological compensation, because the concentrations exhibited by many of these

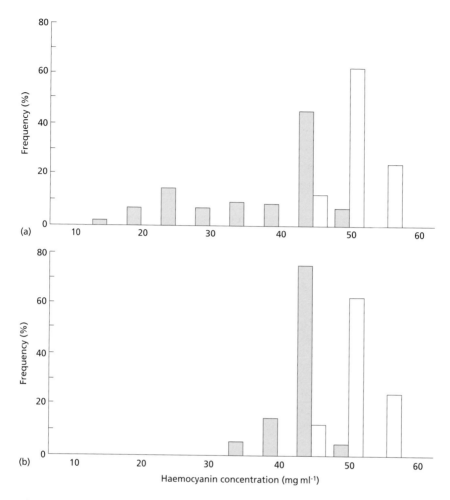

Fig. 3.15 The effect of hypoxia on the concentration of the respiratory pigment, haemocyanin in the haemolymph of the shore crab *Carcinus maenas* (3 days at P_{O_2}, 9.5 kPa; S, 35‰; T, 10°C). Measurements are paired, namely, taken before (shaded) and after (open) hypoxic exposure ($n=155$). (a) All individuals; (b) only the individuals that survived the experimental treatment. (J.I. Spicer, E. Hodgson and T. Traherne, unpublished observation.)

individuals post-treatment are far greater than those exhibited by *any* individuals pretreatment. The degree of compensation exhibited by the survivors can be determined by comparing their haemocyanin concentrations before and after the experimental treatment (Fig. 3.15b). This shows that there is an increase above pretreatment concentrations for the survivors, and that the shape of the distribution of concentrations for the survivors is broadly similar pre- and post-treatment. In other words, both selection and compensation are involved in the population response, but it seems to be that selection alone is responsible for changes in the range of variation and the shift from a skewed to a more symmetrically distributed frequency curve.

3.4 Changes in between-individual variation in the field

Detecting whether or not changes in patterns of between-individual variation in physiological traits occur within populations in the field, and if they do, determining what form they take, is a considerably more difficult exercise than carrying out the laboratory experiments described in the previous section. Perhaps the simplest way to address this is to ask whether individuals in natural populations ever encounter environmental variation beyond their physiological tolerances and capacities (i.e. conditions that could result in a change in the distribution of a physiological trait as a consequence of selection).

Two divergent, if not to say conflicting, views on the match between the physiologies of organisms and the environments which they inhabit permeate the literature. On the one hand, individuals of some species appear to have tolerances and capacities which greatly exceed those required for the conditions which they actually experience. For instance, the Antarctic amphipod *Orchomenella chiliensis* inhabits circumpolar waters where the temperature is both low and roughly constant. And yet in the laboratory this species was able to display a pattern of metabolic compensation to temperature change (−1.8–8°C) that would never be required in its current environment (Armitage 1962). In fact, many species seem to possess much greater tolerances or capacities than they realize in their current environments (e.g. Brown and Feldmeth 1971; Colburn 1988; Ryrholm 1989; Eaton *et al.* 1995; Rombough 1996; Lindburg 1998; Williams and Williams 1998). This notion of individuals having built-in safety margins with regards physiological function and performance is a common one and it has received a reasonable amount of attention (e.g. Wernig *et al.* 1990; Toloza *et al.* 1991; Buddington and Diamond 1992; Hammond and Diamond 1992; Sheard 1992; Lindstedt *et al.* 1994). Diamond (1994) has estimated that most of the safety margins that have been calculated (ratio of physiological capacity to its maximum natural load) have values in the range 1.2–10.

On the other hand, it is also commonly observed that individuals of some species have tolerances and capacities which closely parallel the conditions which they actually experience and that if such tolerances and capacities are exceeded, then mortality should occur. For example, Nordic krill *Meganyctiphanes norvegica* during its diel vertical migration can make excursions into extremely hypoxic waters, for example in fjords with reduced bottom water exchange (Spicer *et al.* in press). While utilization of anaerobic metabolism is critical in allowing them to migrate into hypoxic waters, krill are very close to the limits of their physiological capacities and tolerances at such times. If krill are caged at the depth they inhabit during the day, and so prevented from migrating upwards at dusk, they quickly die (Spicer *et al.* in press). That there are seeming constraints on the physiologies of animals which may be closely matched to what individuals actually experience *in situ* is, again, not novel.

Indeed, there is a growing literature, fuelled by interest in evolutionary constraints, exploring physiological constraints and, in particular, metabolic ceilings (e.g. Peterson *et al.* 1990; Diamond and Hammond 1992; Hammond and Diamond 1992, 1994, 1997; Hammond *et al.* 1994, 1996; Konarzewski and Diamond 1994; Jackson and Diamond 1995) and the design of the respiratory system (Taylor *et al.* 1987, 1996; Weibel *et al.* 1987, 1991, 1992, 1996; Lindstedt *et al.* 1988; Diamond and Hammond 1992; Roberts *et al.* 1996; Vock *et al.* 1996a,b; Weber *et al.* 1996a,b; Gnaiger *et al.* 1998; Hoppeler and Weibel 1998; Suarez 1998). This hypothesis of economic design, in which there is a matching of physiological capacities to each other and to loads, is termed symmorphosis (Taylor and Weibel 1981; Lindstedt and Jones 1987; Diamond and Hammond 1992). In some ways, symmorphosis parallels the proposed close matching of physiological traits to environmental conditions.

Diamond (1994) has suggested that the observed variation in safety margins (i.e. there is evidence for and against symmorphosis; for in-depth discussion see Weibel *et al.* 1998) is best understood in terms of the varying costs and benefits of excess capacity. While it is likely that natural selection will tend to reduce under-utilized physiological capacity because of associated costs, in some circumstances, there may be a greater cost associated with eliminating safety margins. The two analogous views stated here (i.e. that sometimes physiological traits closely match the environment an individual is found in, and at other times, or for other species, they do not) and their relationship to one another, may therefore come into clearer focus if a distinction is made between mortality which is incurred as a result of an extreme climatic event and that which occurs during normal climatic conditions.

3.4.1 Differential mortality and extreme events

Clearly, extreme climatic conditions or events do occur. Such events are often accompanied by what appear to human observers to be dramatic mass mortalities. For example, Kinne (1970) gives a referenced list of marine locations (Bermuda, Denmark, Florida and Texas (USA), Japan, North Sea (Europe)) at which there have been mass mortalities of animal species as a result of cold. Even if mass mortalities do not occur as a result of such events, the physiological function of individuals may be so compromised as to make them more vulnerable to predators, including humans (Horwood and Millner 1998). Populations of the cephalochordate *Branchiostoma lanceolatum* off Heligoland, Germany do not normally experience winter sea temperatures as low as their lower thermal tolerance (see Sect. 4.1), but during the severe winter of 1962–1963 the sea temperature fell to $-1.3°C$, resulting in the mortality of more than 50% of the population (Courtney and Webb 1964). It seems that gradual temperature decreases result in fewer mortalities than do abrupt changes, suggesting that the faster the pace of climatic change, the less likely that the rate of physiological compensation, or alteration, of an individual can keep up. While the actual numbers of marine animals killed by low temperatures can be substantial (e.g. 70–95% of polychaete worm and bivalve mollusc individuals were killed on a sandy beach in Denmark during 2 months of exceptionally cold weather (Blegvad 1929) and 60% of North Sea sole *Solea solea* died in the winter of 1962–1963 (Horwood and Millner 1998)), such losses have yet to be shown to have a significant impact on the persistence of a species in a given area (e.g. Crisp 1964a,b). Unfortunately, few of these studies are quantitative. One notable exception is that of Beukema (1985), who followed winter survival of three sandy beach species, the polychaete worm *Nephtys hombergi*, the cockle *Cerastoderma edule* and the sandmason *Lanice conchilega*. He found that survival could be related to the severity of the winter, and that individuals higher up the shore sustained greater mortalities than those further down (Fig. 3.16).

Such mortalities as a result of low environmental temperatures are not restricted to aquatic environments. For example, during an uncharacteristic 'winter' snowstorm in April 1998, one English hillfarmer reported that overnight 25 of his 110 newborn lambs died of hypothermia; there were no adult sheep mortalities (J.I. Spicer, unpublished observation). Senar and Copete (1995) examined a population of house sparrows *Passer domesticus* at a Mediterranean locality in north-east Spain, and found that the number of freezing days accounted for most of the annual variation in survival rates. These rates declined from 0.4 to 0.5 in mild and normal years to 0.17 in the severe winter of 1984–1985 when January temperatures dropped below 0°C for 12 days. Newton (1998) summarizes other examples for birds. Mortality of pupal stages of the insect *Aphytis* can also be linked with mean monthly minimum temperatures during the winter months (Fig. 3.17).

Kinne (1970) also gives examples of heat death in aquatic (mainly marine) environments, and notes that many animals suffer substantial mortalities at temperatures not greatly in excess of what they normally experience. For example, an increase in temperature to a high of 33°C, as a result of El Niño-related sea water warming, resulted in substantial coral mortality (Brown and Suharson 1990). Droughts as a result of high temperatures may cause mortality in many groups of animals, including birds (Newton 1998).

As well as temperature, there are a number of other extreme natural events which can also result in mass mortalities. For example, Brongersma-Sanders (1957) has reviewed fish mortalities, recorded throughout the globe, and found that, among others, they may result from sudden changes in salinity, algal blooms, severe storms, seaquakes and even volcanic eruptions. Newton (1998) cites cases of mass mortality, occasionally of up to 90% of some populations, among birds resulting from rain storms, tornadoes, hurricanes, blizzards, sea-ice, and (indirectly) fog.

The breadth of phenomena which have been suggested to result in mass mortalities serve to remind us that great caution needs to be exercised in attributing causation to such events, and of the potentially complex links

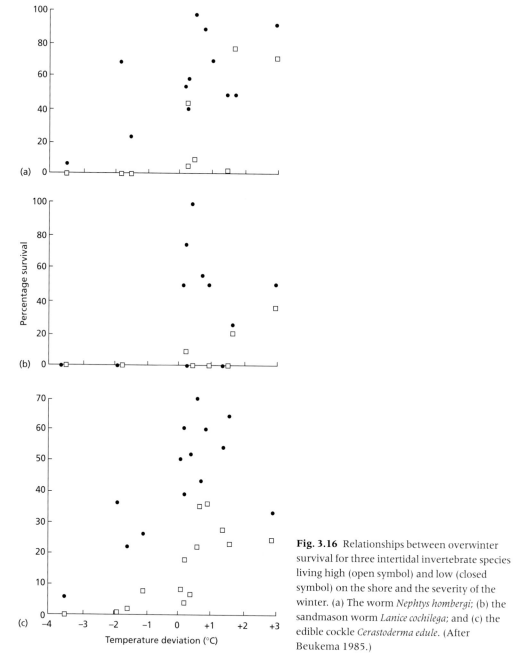

Fig. 3.16 Relationships between overwinter survival for three intertidal invertebrate species living high (open symbol) and low (closed symbol) on the shore and the severity of the winter. (a) The worm *Nephtys hombergi*; (b) the sandmason worm *Lanice cochilega*; and (c) the edible cockle *Cerastoderma edule*. (After Beukema 1985.)

which lead from extreme climatic events to the physiology of an organism. Species in many groups of organisms exhibit local population dynamics which are associated, at least in part, with fluctuations in weather and climate (this is perhaps best documented for insects and birds, e.g. Shelford and Flint 1943; Andrewartha and Birch 1954; Bevan 1976; Cawthorne and Marchant 1980;

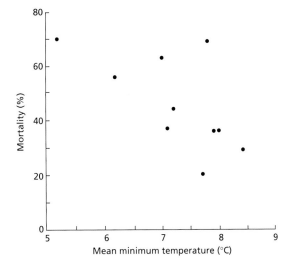

Fig. 3.17 Mortality of *Aphytis* pupae and mean monthly minimum temperatures during the winter months. (After Bursell 1974.)

Pollard 1988; Village 1990; Greenwood and Baillie 1991; Peach *et al.* 1991; Newton *et al.* 1993; Wesolowski 1994; Lawton 1995; Mehlman 1997; Yalden and Pearce-Higgins 1997; Cornell *et al.* 1998; Forchhammer *et al.* 1998; Hawkins and Holyoak 1998; Newton 1998). This does not mean, however, that these dynamics are determined by the physiological tolerances and capacities of individuals, even when they are associated with extreme events. Hirundines (swallows and martins) occasionally suffer dramatic losses, particularly on migration and on their wintering grounds. Torrential rain along the intertropical convergence zone in central Africa has killed large numbers of swallows *Hirundo rustica* and sand martins *Riparia riparia*, and in November 1968 a spell of very cold, wet weather in southern Africa killed many thousands of wintering swallows (Elkins 1995). However, such deaths probably result from starvation rather than from weather conditions *per se*, because the conditions reduce the numbers of insects available. Hirundines are particularly prone to hypothermia, lose heat quickly when it is cold, and huddle together for warmth. Again, this may prevent them from feeding, and result in starvation. Documenting cases of natural die-offs of large mammals, Young (1994) found that many result from starvation following drought.

Returning to the aquatic environment, progressive hypoxia, usually as a consequence of eutrophication or of insufficient water mixing (or both), can result in mass mortality of bottom fauna (Diaz and Rosenberg 1995). This has been best documented for inland or enclosed water bodies such as the Baltic Sea (Baden *et al.* 1990; Rosenberg *et al.* 1990). However, as with all of the above examples, all that are available are rough data on the severity of stress and the resultant mortality. Ideally, one should like to have an example of a change in the frequency distribution of a physiological trait in the field accompanying a climatic challenge.

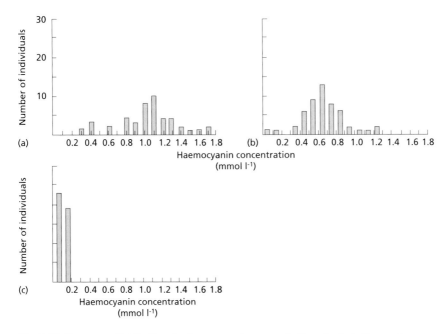

Fig. 3.18 Frequency distributions for haemocyanin concentrations *in situ* in Norwegian lobster *Nephrops norvegicus* in the Baltic Sea (1986–1988). Individuals from (a) normoxic waters; (b) moderately hypoxic waters (43% saturation); and (c) severely hypoxic waters (<10% saturation) ($n=42$–55). (Data kindly supplied by S. Baden-Pihl.)

As part of a long-term study of the Norwegian lobster *Nephrops norvegicus* and exceptional hypoxic events in the Baltic, S. Baden (unpublished; Baden *et al.* 1990) sampled the haemolymph of relatively large numbers of individuals in order to follow what happens to their haemocyanin during natural, but exceptionally severe, hypoxic stress. Reanalysing her raw data, frequency distributions can be constructed for haemocyanin concentration in the haemolymph of individuals of a population experiencing normal oxygen conditions, alongside similar data for those subject to very severe and progressive hypoxia over a number of days or weeks (Fig. 3.18). The normal oxygen individuals, in common with the crabs discussed earlier (Sect. 3.3), show a wide range of haemocyanin concentrations, but unlike the crab example, the frequency distribution is approximately normally distributed (Fig. 3.18a). Those individuals experiencing moderate hypoxia exhibit a reduced range of values and although the frequency distribution is visibly more peaked than is the case for normoxic individuals, the central tendancy is lower (Fig. 3.18b). Under severe hypoxia, the range decreases markedly and there is a pronounced decrease in mean value (Fig. 3.18c). Clearly, in this case, haemocyanin concentration decreases, presumably as lobsters are beyond the point of physiological compensation, and as hypoxia inhibits feeding they are probably using their respiratory pigment as a food source (Hagerman and Baden 1988). Although selection was known to be taking place, as evidenced by the large number of dead individuals brought up by trawl nets, the difference in the pattern of variation cannot be attributed solely to this.

In summary, from time to time there are extreme climatic events which are accompanied by mass mortalities. While these events may be perceived as being relatively rare, this may not in fact be true. Indeed, one should be very wary of statements about the frequencies of supposedly unusual climatic events, without empirical evidence to show that they are indeed such. Weatherhead (1986), in a survey of 380 papers from behavioural, ecological and evolutionary journals, observed that one study in 10 identified the occurrence of at least one unusual event (mostly abiotic), with no simple relationship between the duration of the study and the likelihood of such an event being recognized. He concluded that authors tended to overestimate the importance of some apparently unusual events because they lacked the benefit of the perspective provided by a longer study (most of the studies were only of a few years' duration at the most). It has, for example, been claimed that the occurrence of extremely low temperatures is cyclical (e.g. cold spells every 10 years in waters off Florida and Texas (USA): Storey 1937; Gunter 1957), in which case they often may not be particularly unusual. The brevity of many studies means that, just as is the case for the physiological traits themselves, the temporal variance in climatic events may often be underestimated. This is particularly so if the distributions of climatic variables are strongly skewed (Fig. 3.19).

It remains entirely unclear how important extreme events are to the patterns of between-individual physiological variation which are observed in populations. Presumably, if they occur often enough, they could exert sufficient selection pressure to maintain wide physiological tolerances and capacities (although it is not obvious how frequent this would have to be). Indeed, it is plausible that what appear to be excessively wide tolerances and capacities in relation to the normal range of climatic conditions which individuals experience may in many cases prove to be otherwise when extreme events are considered. It would certainly be interesting to know whether there is any match between the two.

More broadly, if physiological traits are correlated with other facets of the biologies of individuals, then selection on these traits as a result of extreme climatic conditions may also serve to shape the frequencies of these other biological characters in natural populations. For example, survival through drought may in some situations be body size-dependent, perhaps with implications for the subsequent reproductive performance of populations (Jones 1987). Similarly, as a result of severe winter weather there is substantial mortality of cliff swallows *Petrochelidon pyrrhonata* (probably as a result of starvation) with older birds suffering greater mortality than younger ones (Brown and Brown 1998).

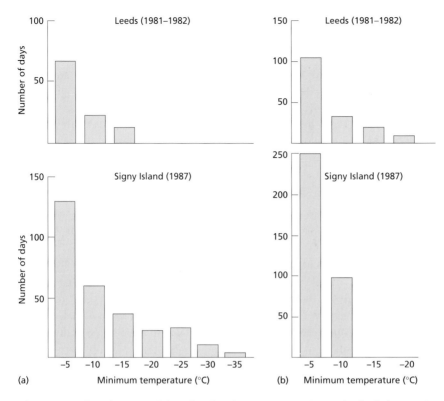

Fig. 3.19 Number of (a) air and (b) soil surface frosts per year, at intervals of 5° below 0°C in Leeds, northern England and at Signy Island, Antarctica. (After Bale 1991.)

In the limit, geographically widespread episodes of catastrophic mortality may explain patterns of local species richness (Wethey 1985). This could have a profound effect on the ecologies of animal populations.

Understanding the net effect of extreme events may be particularly important, given that such events are predicted to increase in frequency with global environmental changes (Wigly 1985; Schneider and Root 1996; Moss 1998).

3.4.2 Population persistence

While mass mortalities have frequently been observed in association with extreme climatic conditions, such events do not seem regularly to serve to drive local populations to extinction. This is not to say that they do not do so, directly or indirectly, on at least some occasions. The California (USA) drought of 1975–1977 was associated with the extinction of several populations of *Euphydryas* butterflies (Ehrlich *et al.* 1980), and unusual spring weather climaxed with a snowstorm in late June in the subalpine area around Gothic, Colorado (USA) caused the extinction of at least one butterfly population through destruction of its host plants (Ehrlich *et al.* 1972). Particularly where local populations are already at the brink of extinction, extreme events may undoubtedly serve to push them over the edge.

But, why do most populations appear to persist through extreme climatic conditions, even when these seem to exceed the physiological tolerances and capacities of the individuals of which they are composed? There may be three reasons. First, it could be that in practice some individuals do not actually experience what are perceived to be extreme events. Many investigators have observed that if allowance is made for behavioural modification and recourse to microclimates (Sect. 3.5.2), mean climatic temperatures rarely exceed the thermal tolerances of even the most temperature-insensitive life stages. The same may be true for the extremes of these temperatures. One of the most temperature-tolerant animals known, the Australian ant *Melophorus bagoti* (CT_{max}=56.7°C), still actively forages during the hottest parts of the day in summer (soil surface >70°C, air temperature at ant height=50°C) by making use of thermal refuges (Christian and Morton 1992; see also Wehner *et al.* 1992 on the Saharan silver ant *Cataglyphis bombycina* for a similar story). Similarly, the carabid beetle *Bembidion andinum*, even though it can be collected from an altitude of 4800 m on Chimborazo, Ecuador, is actually poorly adapted to withstand low temperatures and aridity (Sømme *et al.* 1996). It only survives at this altitude by exploiting sheltered microhabitats under rocks and below vegetation to which it is restricted, venturing out only at certain times during the night when climatic conditions are more favourable. Again, while the sea star *Pisaster ochraceus* of central California (USA) tolerates air temperatures of 21°C for only about 3 hours in the laboratory, it seems to be able to tolerate higher air temperatures than those in the field (Feder 1956). However, the warmest months at the field location at which it has been studied are characterized by heavy fogs and overcast conditions which serve to reduce air temperature. Also, during the warmest months, low tide, when the sea stars are most influenced by air temperatures, usually occurs early in the morning before those temperatures are at a maximum (Hewatt 1937; Feder 1956). This species is further protected from very low temperatures because when these are likely to occur (winter months, early in morning) the tides are neap tides and the starfish remain submerged. Many inhabitants of intertidal pools, such as crabs (Taylor *et al.* 1973), fish (Laming *et al.* 1982) and even shrimp (Taylor and Spicer 1988), will emerse themselves during periods of low oxygen and/or high water temperatures.

Second, populations may persist through events which exceed the measured tolerances and capacities of their constituent individuals, because the ranges of these tolerances and capacities have been underestimated. As discussed earlier (Sect. 3.2.1), if sample sizes are insufficient, good estimates of between-individual variation in physiological traits within populations will not be obtained.

Third, populations may not actually persist through extreme climatic events, but may only appear to do so. This may be especially the case for mobile organisms, and where the loss of a local population may rapidly be followed by immigration of individuals from surviving populations elsewhere.

3.4.3 Normal environmental variation, differential mortality and physiological adjustment

Where cause and effect can be shown, the occurrence of mass mortalities is plainly the consummate demonstration that climatic conditions can exceed physiological tolerances and capacities. However, the deaths of at least some individuals in populations as a consequence of normal climatic conditions is probably a common event. That is, climatic conditions may regularly exceed the physiological tolerances and capacities of a proportion of individuals in many, and perhaps most, populations. Smith (1940) presented good evidence that thermal death of the starfish *Asterias vulgaris* occurs almost every summer in the shallow waters of Malpeque Bay, Prince Edward Island, Canada when the temperature exceeds 25°C. Mortality of the barnacle *Semibalanus* (as *Balanus*) *balanoides* inhabiting (atypically) intertidal rock pools was followed over an 11-week period in the spring/summer. Singletary and Shadlou (1983) record total mortalities between 45 and 80% for each of the pools they examined, with the greatest rates corresponding quite closely with the sharp decreases in oxygen tension that are a regular feature of intertidal pools. In the case of the woodlouse *Oniscus asellus* mentioned earlier (Sect. 3.2.2), there were mortalities *in situ*, correlated with exposure to high climatic temperatures. These temperatures represented the tail end of the frequency distribution of upper thermal tolerances (CT_{max}) determined for individuals from the same population (also see Fig. 3.10; J.I. Spicer, unpublished observation). Although mortality within the woodlouse population was low, the climatic temperatures experienced were still not exceptional.

As with extreme events, determining causal links between mortality and normal climatic conditions may often be difficult, and the paths of causality may have to be carefully disentangled. For instance, the scallop *Placopecten magellanicus* experiences natural periods of low temperature in the Gulf of St Lawrence (Canada) in which it becomes so debilitated that it is very heavily preyed upon by predators (e.g. starfish) that are not so adversely affected by these temperatures (Dickie and Medcof 1963).

Although perhaps more acute in field-based studies than laboratory ones, because of the lack of controlled conditions, in both cases the observation that changes in a physiological trait appear to be associated with environmental change may have to be tempered by the knowledge that many physiological traits exhibit covariance. Thus, even if environmental change is acting directly to change physiology, it may not necessarily be directly affecting the particular trait that is being measured.

3.5 Sources of between-individual variation

What gives rise to between-individual variation in physiological traits within populations? There are four broad, and overlapping, classes of sources of this variation: experimental variation, random developmental noise, differences in the environmental circumstances of individuals and differences in their genetics.

3.5.1 Experimental variation and developmental noise

Bennett (1987a) suggests that perhaps one of the reasons that between-individual variation in physiological traits has been ignored by ecological physiologists is that such variation has not been perceived as being real (i.e. of any biological significance). He then goes on to refute three possible reasons why such a perception has been prevalent. First, the view that extreme values are atypical or abnormal and do not reflect the true response of most individuals is criticised as there is no a priori reason why this should be so.

Second, he addresses the objection that between-individual variation is due to instrumentation or procedural error. Inevitably, some component of the physiological variation which is documented between individuals in a population must result from measurement error. This may, for example, be simply because it is impossible to perfectly replicate the way in which a given trait is measured for even two individuals. None the less, the degree of variation measured in many physiological traits, when contrasted with the accuracy of the methods employed to obtain those measurements, is such that it is plain that measurement error contributes little to this variation (e.g. Fig. 3.20). If such error is a problem for the study of physiological variation, then it is also a problem for experimental studies in physiology more generally. Take, for example, experiments in which temperature measurements from taxidermic mounts are used in estimating metabolic heat production in endotherms. The validity of current practices was tested by Walsberg and Wolf (1996), although their findings have since been contested by Larochelle (1998). Walsberg and Wolf (1996) claim that, at least in some cases, predictions closely matched live animal responses, while others, using data from mounts, produced errors of up to 76% (which could equal the variation produced by different experimental treatments (8.5°C) in the estimate of standard operative temperature).

Third, Bennett (1987a) considered the perception that the physiological variation encountered, if it is real, is both random and unrepeatable. Even individuals that are genetically very similar to one another, when raised in as uniform an environment as possible, often display marked between-individual variation that cannot readily be attributable to experimental error. Such uncontrollable individual variation is most likely due to random internal

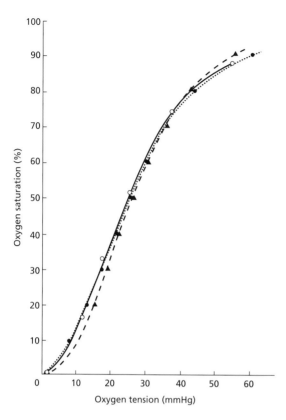

Fig. 3.20 Oxygen-binding curves constructed for human blood (at 37°C, corrected to pH = 7.4) by three different sets of investigators, at different times and each using different techniques. Different symbols indicate different data sets. (After Edwards and Martin 1966, with permission from the American Physiological Society.)

events that take place during development and so is referred to as developmental noise (Maynard Smith 1989, p. 94). The extent to which developmental noise alters the stability of developing physiological traits has not attracted much attention from ecological physiologists. Most of the emphasis has been on the use of developmental noise by geneticists as a measure of the developmental stability of an individual, and analysis of the resultant fluctuating asymmetry and its correlation with genetic diversity, environmental stress and fitness (Scheiner *et al.* 1991; Whitlock 1996; Møller 1997; Ostbye *et al.* 1997; Vollestad and Hindar 1997; Clarke 1998). Once again, the magnitude of between-individual variation in some physiological traits makes it unlikely that this derives from developmental noise alone. However, that developmental noise is, at least in some cases, likely to be a significant factor is evidenced by the fact that attempts to reduce variation in toxicity tests by using genetically inbred strains have met with mixed results (Calow 1996). Indeed, lack of between laboratory standardization, resulting in experimental error and developmental noise, have been confounding factors in the establishment of the generic and repeatable tests required for legislative purposes (e.g. Maltby *et al.* 1987; Aldrich 1991; Soares *et al.* 1992; Forbes and Forbes 1994; Calow 1996; Engel and Vaughan 1996).

Experimental error and developmental noise while in many circumstances they are unlikely to be the dominant sources of physiological variation, may in some cases be potential problems. If so, what is important is that studies correctly estimate these sources of variation.

3.5.2 Individual circumstances

Within any given population, individuals will share the same broad environmental circumstances, simply through their co-occurrence in time and space. However, this may only be true in a gross sense. The precise environmental conditions under which each individual has lived may still differ quite markedly. This difference has the potential to be a major source of between-individual variation in expressed physiological traits. Here, within- and between-individual variation are almost inextricably intertwined (particularly evident in the section dealing with behaviour, Sect. 2.5); at its most extreme, between-individual variation at any moment in real time is constructed from the temporal variation of the within-individual physiologies of a number of individuals (different sizes, ages, developmental stages and environmental histories). Within-individual variation has been discussed in the previous chapter and this material will not be repeated again here. However, it is necessary briefly to explore the repeatability of measurements of between-individual variation through time, the role of microclimate or micro-environment, and variations in tissue physiology of individuals as a possible explanation of between-individual physiological variation.

Repeatability of between-individual differences

For some physiological traits, the repeatability of between-individual variation would appear to be good. For example, this is true of between-year comparisons of the maximum sprint speeds of individual lizards of the species *Sceloporus merriami* (see Fig. 3.21) and between-year comparisons in the responses to hypoxia of individual edible crabs *Cancer pagurus* (Aldrich and Regnault 1990). While Hayes *et al.* (1998) also found that measurements of evaporative water loss and metabolic rate were highly repeatable for Merriam's kangaroo rat *Dipodomys merriami*, they note that in reviewing the literature such repeatability is not always observed. Recent work on metabolic rates in birds and mammals has found that within-individual variation was greater than between-individual variation (Speakman *et al.* 1994; Berteaux *et al.* 1996; Scott *et al.* 1996). For example, for repeated measures (two to six per individual over a 42-day period) of daily field metabolic rates of 11 individual meadow voles *Microtus pennsylvaticus*, after correcting for body mass, 63% of the variation could be accounted for by within-individual differences, with only 27.6% being accounted for by between-individual differences (9.4% was attributable to measurement error) (Berteaux *et al.* 1996). Such poor

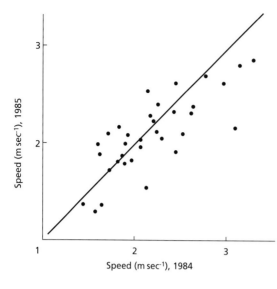

Fig. 3.21 Comparison of maximum speeds in 1984 and 1985 for individual lizards *Sceloporus merriami*. Solid line indicates equality. (After Huey and Dunham 1987, with permission from the Editor of *Evolution*.)

repeatability of mammal field metabolic rates would seem to justify the approach of earlier investigators who have worked only with mean values of this variable, regarding within- and between-individual variation as noise which merely diminishes the capacity of statistical tests to detect between-group differences (Berteaux *et al.* 1996). This said, simple generalizations about the repeatability of measures for different groups of organisms would seem hazardous. Chappell *et al.* (1995) observe that the repeatability of maximum rates of oxygen uptake (after correction for body mass) in wild populations of Belding's ground squirrel *Spermophilus beldingi* were extremely good over a 2-h interval, although repeatability did decline slightly for between-year measurements. Interestingly, the aerobic performance in red junglefowl *Gallus gallus* was, in adults, repeatable up to 180 days, but there was no repeatability for birds tested as chicks and then again as adults (Chappell *et al.* 1996).

These last two observations hint at a more general conclusion, namely that the extent to which within- and between-individual variation are different is context-specific. In particular, it would appear that, with some exceptions, repeatability is greatest when measured over short intervals and declines with time. Given the extent of the changes of physiological regulations and functions with time, discussed in the previous chapter, such a conclusion is perhaps not surprising.

Environmental heterogeneity and microenvironment

The circumstances in which individuals live are almost inevitably much more heterogeneous in the field than in the laboratory. As a result, one might expect that between-individual physiological variation would be greater in field than in laboratory populations. Thus, between-individual variation in cold toler-

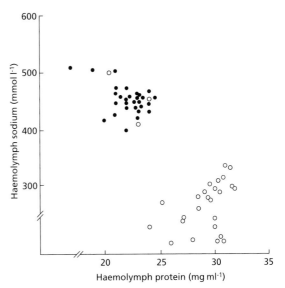

Fig. 3.22 Concentration of protein as a function of sodium concentration in the haemolymph of individual beachfleas *Orchestia gammarellus* freshly collected above (open symbol) and below (closed symbol) the high water mark. (After Spicer and Taylor 1987, with permission from Elsevier Science.)

ance of field-collected populations of the cryptostigmatid mite *Alaskozetes antarcticus* is large compared with individuals kept in the laboratory (Cannon 1987). Similarly, there is less variation in the physiology of painted turtle eggs in the laboratory than encountered in the field (Ratterman and Ackerman 1989). There is considerable variation in the concentrations of sodium and protein in the haemolymph of the beachflea *Orchestia gammarellus* (Spicer and Taylor 1987). Haemolymph from individuals collected from above the high water mark has lower sodium but greater protein concentrations than in individuals collected from below the high water mark (Fig. 3.22). The difference is due to acclimatization and can be accounted for by the fact that individuals living above high water are exposed to considerably more freshwater run-off (and therefore less saline conditions) than those below the high water mark. Such differences can be generated in the laboratory merely by exposing individuals to a range of different salinities. For individuals of some species, differences in environment are generated as a result of biotic factors. For example, the microenvironment of individual salamander *Ambystoma maculatum* eggs is determined by the position of the egg in the egg mass, as well as by differences in external environmental factors (Fig. 3.23).

The relative contribution of individual circumstances to overall between-individual variation in physiological traits within a population is likely to be variable. The more diverse the circumstances in which individuals live, the more substantial this contribution is likely to be. In such cases, environmental heterogeneity can be considered a source of physiological variation. However, even in highly heterogeneous environments, the influence of climatic or environmental influences may be ameliorated by recourse to particular microhabitats.

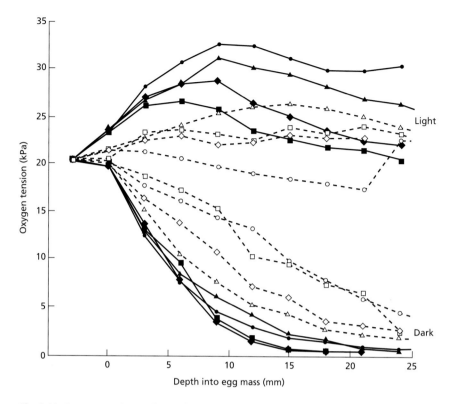

Fig. 3.23 Oxygen tension gradients through four egg masses of the salamander *Ambystoma maculatum* (each represented by a different symbol) at developmental Stages 29–33, under light and dark conditions. (After Pinder and Friet 1994, with permission from the Company of Biologists Ltd.)

Individuals of a terrestrial population may be found in one of any number of different environmental niches, the characteristics of which may bear little relationship to general environmental measurements made for an area. These have been termed microclimates (Waterhouse 1950; Unwin 1980; Rosenberg 1983) or microenvironments (Geiger 1965; Monteith 1973). They can be characterized by fluctuations in environmental factors, such as temperature, humidity and windspeed, and these fluctuations can be either more or less pronounced than those embodied in information normally collected as a source of climate data (Smith 1954; Cloudsley-Thompson 1962; Willmer 1981). To illustrate the type and magnitude of extreme climatic fluctuations that may be encountered at any one location, we have figured as an example how microclimatic conditions vary both spatially and temporally in the Namib desert for 20 h over 2 days in one summer (Fig. 3.24a–c). It can be seen (Fig. 3.24c) that while the body temperatures of six individual beetles are quite variable (30–41.9°C), due to some degree of habitat selection, they are not nearly as variable as the environmental temperatures (30–62.3°C). The important consequences of such habitat selection can be seen at their most

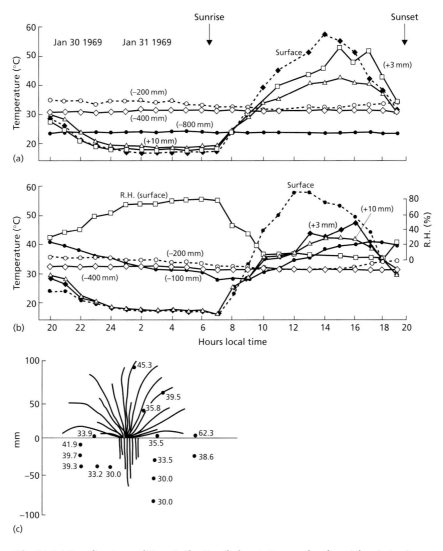

Fig. 3.24 Microclimate conditions in the Namib desert. Temporal and spatial variation in temperature above and below the surface in summer for (a) dune and (b) dune slope. (c) Spatial variation in air and soil temperatures (right) and body temperatures for white beetles *Onymacris langi* (left) at midday. R.H., relative humidity. (After Willmer 1981, with permission from Academic Press.)

extreme in the case of garter snakes *Thamnophis elegans* living beneath rocks of different thickness (Fig. 3.25). While individuals living beneath thick rocks are still exposed to temperature fluctuations, such fluctuations are never so extreme that they exceed their physiological capacities and tolerances. This would not be the case under thin rocks, where the range of temperatures available around midday all exceed the thermal tolerance of an individual living there (Fig. 3.25). Recourse to microclimates is seen as being critical to the survival of many desert (e.g. Cloudsley-Thompson 1991; Wolf *et al.*

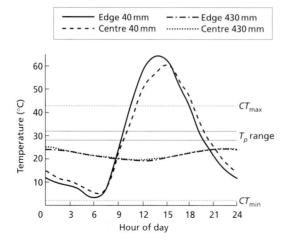

Fig. 3.25 Spatial and temporal variation in temperature beneath thick (430 mm) and thin (40 mm) rocks at Snakehenge, California (USA). Superimposed are the upper and lower temperatures at which the garter snakes *Thamnophis elegans* which inhabit this region lose their righting response (CT_{max} and CT_{min}, respectively) and their preferred temperature range (T_p). (After Huey 1991, with permission from The University of Chicago Press.)

1996) and high latitude species (e.g. Davenport 1992). However, it may also be critical in most environments. It is just that usually environments typified by 'extreme' conditions are, perhaps erroneously, perceived as more interesting objects of study.

Generally speaking, one of the main features of the open seas and oceans is that they are relatively homogeneous and so recourse to microenvironments by their inhabitants, it has been argued, is much less relevant when compared with terrestrial systems (e.g. Bartholemew 1958). That said, different micro-environments do exist in the marine environment, in and around oxygen minimum layers (Childress 1995); hydrothermal vents (e.g. Johnson *et al.* 1988); sea ice (S.L. Chown, personal communication); in semi-enclosed or shallow coastal waters, and in intertidal pools (Fig. 3.26); and also in fresh-water systems (Fig. 3.27). While the range and rate of temperature change may be less pronounced in aquatic environments, mainly as the thermal conductivity of water is so much greater than that of air, this is not the case for gases, such as oxygen or carbon dioxide. These respiratory gases can vary dramatically, often over very short distances, particularly in small or highly stratified bodies of water, such as isolated fresh and seawater pools.

While their exploitation can result in amelioration of between-individual physiological variation, the actual outcome will be determined by the availability of suitable microenvironments (cf. Beck 1997). If these are limiting, then it might be expected that some individuals will be excluded from them, thereby increasing the physiological variation in that population.

Sub-individual variation as a source of between-individual variation

There is often between-individual variation in tissue or organ size. This variation could potentially account for between-individual differences in physi-

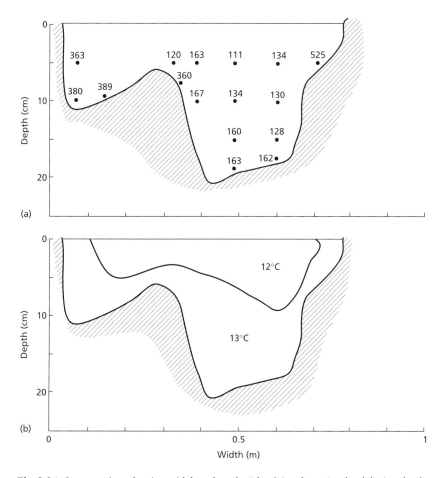

Fig. 3.26 Cross-section of an intertidal pool on the Isle of Cumbrae, Scotland during the day showing variations in (a) oxygen tension as spot measurements and (b) temperature. (After Morris and Taylor 1983, with permission from Academic Press.)

ology. For example, similar-sized individuals may exhibit a range of sizes of metabolically active tissues or organs (which contribute disproportionately to overall metabolism), and so possess markedly different metabolic rates. Furthermore, the upkeep of such metabolically active tissue is likely to entail maintenance costs, thereby further elevating metabolic rate. Working with small mammals, Konarzewski and Diamond (1994, 1995) found that high food intakes and elevated resting rates of metabolism are characteristic of individuals with large kidneys, livers, hearts and intestines. Speakman and McQueenie (1996) found that individuals with high basal metabolic rates also had large digestive organs. Some support for the positive relationship between metabolism and organ size was also provided by Koteja (1996) for mammals, and by Garland and coworkers (Garland 1984; Garland and Else 1987) for lizards. In the tree swallow *Tachycineta bicolor*, however, while differences in

Transverse section   Horizontal section

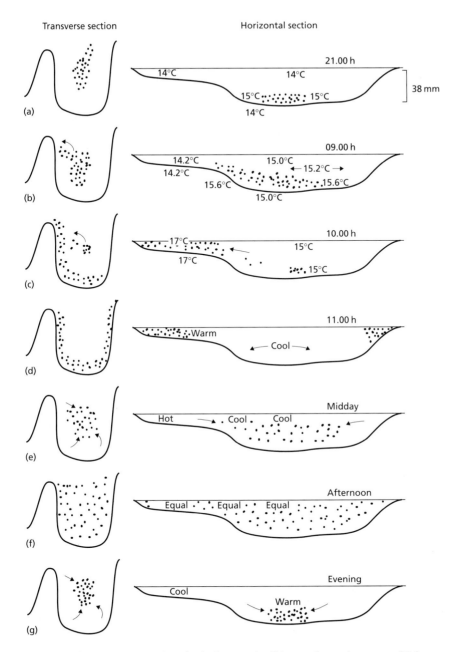

Fig. 3.27 Changes in aggregation of tadpoles *Rana boylii* in a pool over the course of 2 days in late June (Del Norte County, California, USA). (a) During the night and (b) early in the morning tadpoles are concentrated at the bottom and in the middle of the pond (14–15°C). During the day they are able to exploit changing water temperatures by movement to various microhabitats (c–g), although by evening they are reaggregated at the bottom and in the middle of the pond (g). (After Brattstrom 1962.)

the sizes of kidney and small intestine accounted for 21% of the variation in metabolism, individuals with high resting metabolic rates had larger kidneys but smaller intestines and pectoral muscles than did individuals with low resting metabolic rates (Burness *et al.* 1998). So, within a population of a species, relationships between metabolic rate and organ mass do exist but they may not be consistent. It should be noted that although such variations in tissue/organ mass are treated here as environmentally determined (e.g. as a result of diet and energy demands; Daan *et al.* 1989; Gaunt *et al.* 1990; Dykstra and Karasov 1992), there is also some evidence for a genetic component (Konarzewski and Diamond 1995).

3.5.3 Genetic differentiation

Between-individual variation in physiological traits may have a genetic basis. Unfortunately, in contrast to the link between genetic and morphological/ ecological diversity (Sect. 1.3), the link between genetic and physiological diversity has been comparatively poorly studied (Watt 1985; Koehn 1987; Powers 1987). On one level, finding between-individual differences that may be genetically determined is comparatively easy. However, examining the genetics of physiological differentiation between individuals is considerably more involved and more difficult to handle. As neither of the present authors are geneticists, an attempt to redress this imbalance will not be made here. We will outline some examples where genetically determined between-individual differences in physiology have been observed, and consider the importance of such differences in contributing to overall between-individual physiological variation.

Individuals belonging to the same population, and perhaps living literally centimetres away from each other, can show pronounced variation in physiology as a result of exploiting different microhabitats and thus experiencing different circumstances (Sect. 3.5.2). However, in some cases such variation may result because groups of those individuals belong to genetically distinct lines or clones. For example, crevices at different levels on rocky shores in North Wales support populations of the small intertidal bivalve *Lasaea rubra*. Each crevice population consists of different proportions of three genetically distinct inbred lines (Tyler-Walters and Davenport 1990). Between-individual variation in upper median lethal temperature for a given population is determined by the relative proportions of the three lines which are present (Fig. 3.28). This in turn leads to differences in upper median lethal temperatures of populations on progressively lower sections of the shore, and consequently gives rise to between-population differences (cf. Sect. 4.2.4).

It would be interesting to know just how prevalent the phenomenon of the co-occurrence of morphologically similar, yet genetically distinct, individuals actually is in the field. Clonal genotypes with different physiologies have been examined in a number of species of sea anemone (Shick and Dowse

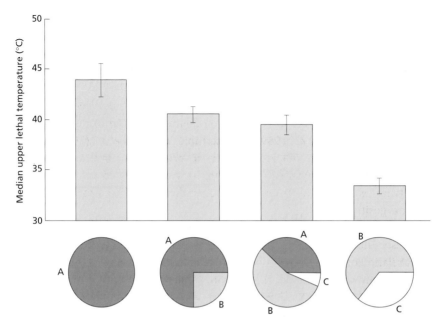

Fig. 3.28 Median upper lethal temperatures for high-shore crevice populations consisting of different proportions of three genetically distinct inbred lines (A, B and C) of the intertidal bivalve *Lasaea rubra*. (Data from Tyler-Walters and Davenport 1990.)

1985; McManus *et al.* 1997). Six co-occurring colour morphs of the brittlestar *Amphipholis squamata* show differences in the intensity of bioluminescence and in the neurophysiological control of light production (different cholinergic receptors and neurohormones). While this between-individual physiological variation is presumably genetic, the intensity of luminescence was noted to be greater in brooding compared with non-brooding individuals, suggesting the possibility of a non-genetic component (Deheyn *et al.* 1997). The locomotory performance of a number of lizard species has been relatively well studied by ecological physiologists. While such performance is highly variable among individuals, this variation is repeatable and has been shown to have a genetic component (e.g. Bennett and Huey 1990). Similarly, while the expression of heat shock proteins is largely under the influence of environmental temperature, even this is under genetic control (Krebs and Feder 1997a).

Some effort has gone into examining allozyme variation between individuals, and into trying to correlate genetic variation with physiological variation (Johnson 1979). While correlations are relatively easy to detect, exactly what they mean is not always clear. For example, although characterized by relatively low levels of genetic variation, the honey bee *Apis mellifera* is found to possess a number of polymorphic enzymes, including cytoplasmic (or soluble) malate dehydrogenase (s-MDH) (Coelho and Mitton 1988). Both slow (*S*) and fast (*F*) alleles are present in one population that has been studied, and the mass-specific maximal oxygen uptake (VO_{2max}) of flight muscles of *FF* homozygote workers is significantly greater than that of *SS* homozygote workers. In drones, this pattern is reversed. Unfortunately, the role, if any, of s-MDH in variation in VO_{2max}, or at least what s-MDH polymorphism and VO_{2max} covary with, is unclear.

As well as focusing on individual enzymes, attempts have been made to relate overall genetic diversity to physiological traits (e.g. Garton 1984; Mitton *et al.* 1986; Hawkins *et al.* 1989). For example, Hawkins *et al.* (1989) found that individuals of the blue mussel *Mytilus edulis* characterized by a high degree of heterozygosity survive exposure to water-borne copper longer and possess lower protein turnover rates than individuals with a low degree of heterozygosity.

In these examples, variation in a particular enzyme or suite of enzymes has been used almost as an index of genetic diversity and, as stated above, the exact relation of such diversity to precise function is not clear. However, there are also cases where variation in allozymes known to be critical to particular physiological processes has been studied, hand-in-hand with studies in physiological variation. For example, the genetics of the enzyme glutamate-pyruvate transaminase (GPT) has been studied in the intertidal copepod *Tigriopus californicus*. Adults with fast migrating allozyme (homozygous *FF* or heterozygous *FS*) accumulate free amino acids, especially alanine, in a hyper-osmotic medium. Larvae that are *SS* homozygous show higher mortality under hyperosmotic stress (and are also incidentally less heat tolerant) than those that are *FF* homozygous (Burton and Feldman 1983). Finally, it should be noted that enzyme polymorphism may, via flux through metabolic pathways, affect developmental stability (see Sect. 3.5.1; Mitton 1995).

Genetic diversity plainly does contribute substantially to between-individual variation in physiological traits in certain cases, although it has not attracted the detailed consideration which would seem its due. Studies in which genetic diversity could not be shown to be related to between-individual differences in physiology seem more difficult to locate. None the less, it would be surprising if such cases did not exist, and this may perhaps represent an example of the so-called 'file drawer' problem, an under-reporting of studies which failed to find evidence of the patterns which were being sought (Rosenthal 1979; Csada *et al.* 1996).

3.6 The link to fitness?

Darwin (1859) first observed that between-individual variation must affect fitness for evolution to occur by natural selection. Much subsequent work has focused on determining the relations between fitness and variation in morphological, behavioural and life history traits (e.g. Endler 1986; Lessells 1991). Relations between fitness and variation in physiological traits have, in contrast, remained relatively little studied, although the gradual acceptance of the fact that physiological systems and processes may not be optimized is beginning to make its way into physiological textbooks. Randall *et al.* (1997, p. 5) observe that, given the precision of some other known physiological controlling systems, it is conceivable that a more precise control system for temperature could exist than is found at present in most mammal species. However, they go on to note that such a system has not been fixed by selection, indicating that a 1–2°C temperature range is tolerable, and must be good enough for survival.

Of the studies carried out explicitly to establish and then examine the relationship between fitness and physiological variation, not all have been successful in establishing such a link. This point is made by Endler (1986), Feder (1987a) and Kingsolver and Huey (1998). For example, Walton (1988) combined replicated measurements of aerobic metabolism and of locomotory performance with observations of behaviour under natural conditions to determine whether individual variation in activity metabolism was related to variation in locomotory performance and foraging behaviour among Fowler's toads *Bufo woodhousii fowleri*. Although individual toads differed significantly in activity metabolism and locomotory performance, individual variation in metabolism did not explain variation in locomotory performance. In the field, toads that hopped frequently and moved greater distances also struck more frequently at prey and consumed more beetles than toads that were sedentary. However, metabolism during activity and locomotory performance, as measured in the laboratory, were either unrelated to behaviour in the field or showed relationships the opposite of those expected. Similarly, links between individual variation in metabolic rate and life history traits have, to date, proved elusive (e.g. Hayes *et al.* 1992). The possible reasons for the difficulties associated with relating physiological variation (or any biological variation for that matter) to fitness are many and complex; they are presented in detail by Endler (1986). Fortunately, such difficulties have not prevented some workers from exploring the link between ecological physiology and (components of) fitness, and sometimes elegantly so (Kingsolver and Watt 1983; Kingsolver 1985, 1989, 1995a,b, 1996; Bennett 1987a; Huey and Kingsolver 1989; Bennett and Huey 1990; Kingsolver and Wiernasz 1991; Kingsolver *et al.* 1993; Kingsolver and Huey 1998).

Perhaps for the majority of studies where the fitness implications of particular patterns of physiological variation have been proposed, such suggestions have not been tested directly. Sometimes proposed explanations take the form of merely thinking up adaptive reasons (despite the critique of the adaptionist approach by Gould and Lewontin 1979), but in other cases the links made seem more credible. For example, in Chapter 2 it was observed that the body temperature of free-living African or Jackass penguins *Spheniscus demersus* varies with (foraging) activity (Sect. 2.5). However, instead of regarding this as merely an environmentally mediated disturbance in a physiological regulation, such variation, it has been suggested, may actually be beneficial to individuals of this species. For typical dive and rest durations, birds with flexible body temperatures show greater foraging efficiencies (approx. 10%) than birds with fixed body temperatures (Wilson and Grémillet 1996). Also in the previous chapter note was made of critical windows for environmentally modifying the physiological itinerary associated with sex determination in reptile embryos (Sect. 2.4.3). In this connection, Burger (1998) shows that pine snakes *Pituophis melanoleucus* hatching from nests with low incubation temperatures might be less able to avoid predators and find shade at higher environmental temperatures than individuals from nests characterized by higher incubation temperatures; if so, the former could be expected to show lower survival in the wild. Finally, Watkins (1996) found that tadpoles of the Pacific tree frog *Pseudacris regilla* that escaped predation by the garter snake *Thamnophis sirtalis* swim over one and a half times as fast and nearly four times as evasively as those that are captured. Consequently it was suggested that, as snake predation on tadpoles is common, burst swimming speed may be subject to predator-mediated selection in nature. Despite the attractiveness of such explanations they are, as yet, still untested.

Feder (1987a) in his consideration of physiological diversity posed the question of what the relationship is between physiological diversity and fitness. He figured two possible extreme forms of the relationship between a physiological trait and fitness (Fig. 3.29). The first assumes that physiological traits are under strong stabilizing selection which results in an extremely leptokurtic (tight-peak) distribution. Here, any deviation from the physiological optimum will result in a pronounced reduction in fitness. In the second form,

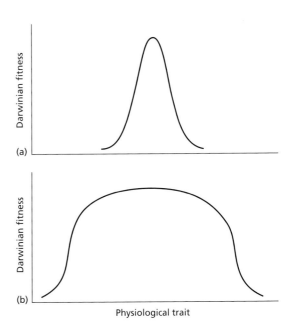

Fig. 3.29 Two extreme forms of the possible relationship between physiological variation and fitness. (After Feder 1987a, with permission from Cambridge University Press.)

the breadth of physiological variation that can be tolerated before there is any marked loss of fitness is extremely wide, resulting in a platykurtic frequency distribution (i.e. there is a pronounced plateau). However, as Feder (1987a) points out, it is not inconceivable that physiological variation in particular traits may have to be so wide before fitness is affected that detecting, let alone defining, such relationships will be difficult if not impossible.

Unfortunately, few experimental studies have tested Feder's ideas. In a recent investigation of the waterflea *Daphnia magna* (measured at any given developmental stage) the relationship between fitness, measured as lifetime reproductive output, and heart rate is not a tight one (Fig. 3.30a). Only very large deviations in heart rate are associated with marked changes in fitness and there is a distinct plateau to the relationship, which is similar to that of Feder's second diagram (Fig. 3.29b). A frequency distribution for heart rate shows a strong central tendency, with most individuals displaying a rate of approximately between 370 and 400 beats min^{-1}, and a strong left skew (Fig. 3.30b). Presented in this way, these data are firmly at odds with the notion that minor variations in physiology will have important consequences for fitness. Instead, in this case at least, it seems that relations between fitness and physiological traits may be observed only when individuals exhibiting wide variation in the latter are examined.

The relationship between physiological variation and fitness need not necessarily be dome-shaped, it could also be linear. A considerable amount of attention has been given to the relationship between the locomotor performance of reptiles and indices of fitness, particularly for individuals from real

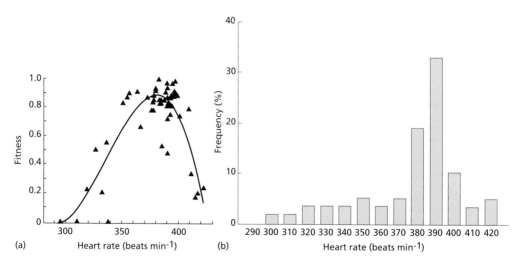

Fig. 3.30 Heart rate of the waterflea *Daphnia magna*. (a) Relationship between heart rate and fitness (estimated from lifetime reproductive output) for individuals of the same developmental stage and mass. (b) Frequency distribution of heart rate for those individuals ($n=57$). (After J.I. Spicer and K.J. Gaston, unpublished.)

populations in the field (Bennett and Huey 1990). In a 3-year study of physiological performance (speed, exertion and endurance) of a population of garter
snakes *Thamnophis sirtalis*, it was found that in their second year of life, speed
and exertion were related to survival (analysed using size-corrected residuals,
as both locomotor traits and survivorship are related to body size) and for individuals older than 2 years' endurance was also related to survival (Bennett
and Huey 1990). Such relationships could not be detected in a similar study of
fence lizards *Sceloporus occidentalis*, which Bennett and Huey (1990) suggest
may be because the population studied was close to the northern limits of its
range, where there are fewer predators. They go on to note that although the
field studies they present are hypothesis driven, such exploratory approaches,
although necessary, are only a first step in studies of fitness and are a prelude
to more manipulative work.

The possibility of genetically modifying animals will, in the foreseeable
future, provide an invaluable tool for probing the fitness implications of
physiological variation. It is now possible to turn on, or block, the expression
of particular genes or gene groups that code for known physiological traits.
Such techniques are used to generate particular patterns of variation in
targeted physiological traits but without necessarily affecting other traits
that often covary when this variation is generated using environmental
factors. Genetically modified animals are currently being used to address
fundamental physiological questions in areas such as renal function (e.g.
Kopp and Klotman 1995; Fukamizu and Murakami 1997); haemodynamics
(e.g. Barbee *et al.* 1994; Thompson *et al.* 1995); nutrition (e.g. Knapp and
Kopchick 1994; Morin and Eckel 1997); brain and nervous function (e.g.
Lathe and Morris 1994; Fritz and Robertson 1996); vision (e.g. Feiler *et al.*
1992); cardiovascular function (e.g. Brosnan and Mullins 1993); thermal
tolerance (e.g. Wang *et al.* 1995); and trace metal regulation (e.g. De Lisle *et al.*
1996).

The potential for use of genetic manipulation to test fitness consequences
or implications of particular physiological traits is to some extent already being
realized (Feder and Block 1991). For example, Feder and his coworkers have
examined natural and environmentally induced variation in a heat shock
protein expressed in fruit flies, and from such studies have drawn inferences
on the effects of such variation on components of fitness (Feder *et al.* 1997;
Krebs and Feder 1997a,b). As Feder (1996) has pointed out, the only way to
examine the contribution of one particular trait to fitness is to manipulate that
trait while all other traits (many of which would normally covary with the
trait of interest) are held constant. Consequently, he and his coworkers have
used transgenic techniques to modify the expression of heat shock protein
alone (Feder 1996; Feder *et al.* 1996; Krebs and Feder 1997c). They find that
by producing extra heat shock protein, genetically modified fruit flies sustain
less damage during thermal stress, but if the production is too great there
results a marked decrease in growth, development and survival (Krebs and

Feder 1997c). Using this powerful technique, such fundamental questions that previously could only be addressed by inference can now be answered directly, such as whether it actually matters to what extent (or even whether or not) an animal can express heat shock or respiratory proteins. In the case of the latter, the results of a recent study indicate that the answers to such questions look as if they may pose a serious challenge to current understanding of the ecological and evolutionary significance of physiological traits. Garry *et al.* (1998) found that they were able to genetically modify mice not to have the intracellular respiratory pigment myoglobin. Muscle myoglobin is thought to be important, if not essential, in taking oxygen from the blood and transferring it to the tissues, particularly during times of heightened metabolic demand. And yet these transgenic mice showed no difference in their abilities to meet the metabolic demands of pregnancy or exercise compared with control individuals!

In conclusion, the physiological and ecological literature is replete with adaptive reasons for variation in particular physiological traits. However, rarely have these explanations been the subject of rigorous testing. Where this has been done, the validity of such reasoning has not always been substantiated. Nowhere is this clearer than in the case of what almost seems axiomatic for most biologists, the beneficial nature of acclimation.

Is acclimation beneficial?

It is generally assumed that acclimation confers fitness benefits on the perfor-mance of animals; 'acclimation to a particular environment gives an organism a performance advantage in that environment over another organism that has not had the opportunity to acclimate to that particular environment' (Leroi *et al.* 1994a,b). This intuitively attractive idea is termed the 'beneficial acclimation hypothesis' (Leroi *et al.* 1994a,b; Crill *et al.* 1996; Huey and Berrigan 1996; Bennett and Lenski 1997). Unfortunately, the hypothesis has not received the scrutiny it deserves (Hoffmann 1995; Huey and Berrigan 1996; Kingsolver and Huey 1998), given its central importance for understanding of the ways in which individuals respond to environmental challenge. Only a handful of recent studies has sought to test the beneficial acclimation hypothesis for animals (and bacteria), by examining the relationship between the thermal conditions individuals are maintained under for a given period of time and subsequent performance in different thermal envi-ronments. In the majority of these cases the hypothesis has been rejected (Leroi *et al.* 1994a,b; Zamudio *et al.* 1995; Huey and Berrigan 1996; Bennett and Lenski 1997; Gibbs *et al.* 1998; Temple and Johnston 1998), although this has not invariably been so (Leroi *et al.* 1994a,b; Bennett and Lenski 1997; Scott *et al.* 1997; Temple and Johnston 1998). Thus, for example, Bennett and Lenski (1997) found that, contrary to prediction, a lineage of *Escherichia coli* evolutionarily adapted to 37°C and maintained at 37°C has a higher fitness at 32°C than do bacteria maintained at 32°C. Indeed, they found that the hypothesis is not supported in nearly half of the comparisons made. Likewise, Huey and Berrigan (1996), reanalysing data from Zwaan *et al.* (1992), found that contrary to prediction the longevity of fruit-flies *Drosophila melanogaster* is greater following development at intermediate temperatures when adult flies are living at cool, intermediate or warm temperatures, and that development at cool or warm temperatures conveyed no advantage.

The studies of the beneficial acclimation hypothesis that have been performed to date have almost universally been elegant and incisive. They suggest that explanations for patterns of acclimation must be based on alternative evolutionary hypotheses (e.g. random drift, adaptation to past climates, gene swamping; Huey and Berrigan 1996), or that acclimation may not always be adaptive (Hoffmann 1995). If such conclusions were to generalize they would be very significant indeed, simply because acclimation is such a widespread phenomenon. However, and perhaps inevitably given their novelty, empirical studies of the beneficial acclimation hypothesis suffer from a number of significant limitations (often freely admitted by their authors), which require resolution before such generalization would begin to look secure.

First, those studies of the beneficial acclimation hypothesis which have been conducted to date concern a very narrow range of taxa. Indeed, they almost universally concern *E. coli* and *D. melanogaster* (but see Temple and Johnston 1998). This is not uncommon in evolutionary studies and, following the Krogh principle (see Sect. 1.3), these organisms provide an obvious point of departure. Second, while work on *E. coli* has utilized direct measures of Darwinian fitness (e.g. Leroi *et al.* 1994a,b), the measures of performance employed in other studies have often been rather more indirect expressions of fitness (e.g. territorial success, longevity, water loss, escape performance). In some cases, these latter measures are probably reasonably close correlates of Darwinian fitness, but in others it is less obvious that this need be so. Resorting to such indirect methods is often a product of necessity, resulting from the practical difficulties associated with more direct measures for particular species. Third, the studies have almost exclusively been conducted in the laboratory. This provides a logical starting point. However, it ignores the complexities of field conditions, and ultimately field-based tests are not just desirable, but in fact are an essential component in testing the ecological significance of physiological responses (Huey and Berrigan 1996). In other words, it is necessary to establish whether the field equivalent of acclimation, acclimatization, is also beneficial; that is, is the 'beneficial acclimatization hypothesis' falsified? Certainly there is no necessary a priori reason why acclimation and acclimatization experiments in general should have the same outcome; indeed, the evidence to hand indicates that the opposite is true (Cossins and Bowler 1987; Gatten *et al.* 1988; Roberts *et al.* 1997; Seddon and Prosser 1997).

In summary, contrary to what might have been expected, there is no overwhelming support for the beneficial acclimation hypothesis. Such a salutary lesson should serve to further caution against forming adaptive explanations, however irresistible they seem.

3.7 Concluding discussion and summary

'The fitting of a given distribution to a set of data should not be an end in itself, but should serve as a guide to the factors underlying such distributions.' [Barnes, 1967, p. 18]

Within populations, there is often pronounced and significant between-individual variation in physiological traits. This is true even for physiological processes which are relatively tightly regulated. The frequency distributions expressing this variation may exhibit a wide variety of shapes, and while approximately normal distributions may not be unusual, it seems likely that they are far from universal. These patterns may be influenced by the extent to which populations are composed of individuals of different ages, body sizes and/or developmental stages, the strength of this influence depending on the degree to which physiological traits change with age, size and development.

All of the defining features of a frequency curve for a given physiological trait (i.e. its spread, central tendency, kurtosis and skew) are open to experimental modification in the laboratory. These changes can come about by selective mortality, by physiological adjustment, or by some combination of the two. In the field, both extreme and normal climatic events may be responsible for altering the frequency distribution of physiological traits, again both by selective mortality and physiological adjustment (or damage), although the former is more pronounced under extreme conditions. The frequency distributions of physiological traits within populations may on the one hand determine the likelihood of the population persisting through extreme events, while on the other hand these distributions may themselves be moulded by these conditions.

Regardless of its pattern, there are three main sources of between-individual variation in physiological traits within populations: experimental error, differences in environmental circumstances and genetics. Experimental error may be important in some investigations, but it is not sufficient to explain most differences between individuals. The implications of between-individual variation in physiological traits for fitness are only beginning to be explored, although from the few studies that have been carried out so far the association may at best be a loose one. Clearly, an understanding of the extent and nature of within- and between-individual variation is important for ecologists and, re-echoing the words of Feder (1987a), we still urgently require carefully executed studies investigating the nature and origin of unconstrained within-population physiological variation, and the effect of climatic factors on such variation.

Chapter 4: Population Differences

'How much of the acclimatisation of species to any particular climate is due to mere habit, and how much to natural selection of varieties having different innate constitutions is an obscure question. I must believe that habit has some influence, but the effects have been largely combined with, and sometimes overmastered by, natural selection.' [Darwin, 1859]

4.1 Introduction

Given the prevalence of between-individual variation in physiological traits within populations and the fact that this variation is influenced by the environment, it seems unlikely (although not impossible) that different populations of the same species will have identical physiological profiles. That is, one would expect there to be between-population physiological variation. So it should come as no surprise that reports of physiological differences between populations are commonplace in the literature; such differences are relatively easy to detect, and thus have been discussed, or referred to, in many reviews and books (e.g. Stauber 1950; Prosser 1955; Vernberg 1962; Vernberg and Vernberg 1964, 1972; Garland and Adolph 1991; Maltby in press). A selection of early studies includes that of Goldschmidt (1918), who found that the physiologies of individuals of the gypsy moth *Lymantria dispar* could be related to the climate under which they occurred, and of Timoféeff-Ressovsky (1940) (both cited in Huxley 1942), who found that individuals of a species of *Drosophila* widespread in Europe differed in temperature resistance depending on exactly where they were collected. Similarly, Whitney (1939) found that mayfly nymphs had different upper thermal tolerances depending on the temperature of the stream from which they were collected. Likewise, Finn (1937), studying two populations of a species of American salamander, observed that they differed markedly in the size of their red blood-corpuscles. Such examples serve to illustrate that there has been an awareness of population differences in physiology for some considerable time.

None the less, there are also counterexamples. That is, there are populations of individuals of the same animal species which possess similar physiologies despite their inhabiting geographically widely separate localities (sometimes referred to as 'static', as opposed to 'labile', responses). The cephalochordate *Branchiostoma lanceolatum* is found in both temperate and tropical waters. When the thermal tolerances of individuals from populations off Naples, Italy were compared in the laboratory with those from off Heligoland, Germany there was no difference (over the range 3–27°C;

Courtney and Webb 1964). Temperature and salinity tolerance of two populations of the hermit crab *Pagurus longicarpus*, one from Massachussetts and the other from South Carolina, USA, were near identical (Young 1991). Likewise, Matthews (1986) found no significant differences in the laboratory in mean critical thermal maximum between 18 populations of a minnow, the red shiner *Notropis lutrensis,* although these populations represented all of the major river systems occupied by the species across an 1100 km north–south span of its range in the American Midwest. There is also work by Russian scientists (cited by Berman and Zhigulskaya 1995) who found no significant population differences in cold resistance for a number of lepidopteran and coccid species.

Most studies which fail to find significant differences in physiological traits between natural populations have been conducted in the laboratory, which may, as shall shortly be seen, have some profound consequences. Few workers have compared the physiological traits of individuals in different populations straight from the field, as has been done by some investigators interested in between-individual variation within populations (Sect. 3.4). One exception is a study of three geographically isolated populations of the shrimp *Palaemon adspersus* taken from fjords in the Baltic. Individuals from these populations did not differ in their standard metabolic rates, although each collecting area (and hence experimental examination) had its own distinct salinity and temperature regime (von Oertzen 1984).

On balance, it seems likely that circumstances in which populations exhibit differences in physiological traits substantially predominate over those in which they do not. The origin of these differences and what determines these two outcomes remains, none the less, an important issue. In this chapter, we examine the different sources of between-population physiological differences (where they exist) and how they can be identified. Some of the patterns detected with respect to clines and the geographical ranges of species are presented and, where possible, the underlying mechanisms are briefly explored. We conclude with a consideration of how physiology may influence or constrain the geographical range of a species. Throughout, many, although by no means all, of the examples which will be drawn on concern physiological traits related to environmental temperature. This reflects the relative predominance of such studies in the literature, but also the probable importance attributed to this environmental factor in determining the distributions of species (e.g. Merriam 1894; Allee 1923; Hutchins 1947; Allee *et al.* 1949; Crowson 1981; Jeffree and Jeffree 1994; Somero 1997).

4.2 The roots of population differences

4.2.1 Things are not always as they appear

While most populations probably show some difference in the expression of physiological traits, this does not necessarily constitute physiological differentiation. It could be that some populations are physiologically different *in situ*, but when individuals are brought back to the laboratory, and kept under similar conditions, any differences disappear. Individuals from these populations would be described as displaying local physiological acclimatization. At the other extreme, some populations show persistent physiological differences even on being maintained in the laboratory for many generations under conditions different from those under which they originally occurred, and would be described as exhibiting local genetic adaptation. Local physiological acclimatization consists of phenotypic adjustment of a physiological response (albeit within prescribed genetic limits), while local genetic adaptation involves selection of mutations.

Even if there are differences in physiological traits between populations as a result of local genetic adaptation, this does not automatically mean that these are of tremendous ecological significance. For example, there are marked differences in the effect of water-borne zinc on osmoregulation in the freshwater amphipod *Gammarus pulex*, from two populations that differ in their metal sensitivity, one a metal-tolerant population that occurs in a stream contaminated by metalliferous material as a result of mining over the last 200 years (Spicer *et al.* 1998). However, these physiological differences are only manifest at concentrations of zinc in excess of those known to affect growth and reproduction in the population from the uncontaminated site, and are greater than those occurring at the metal-contaminated site. This does not necessarily mean that the physiological differences are unimportant, as not all of the constraints acting on the physiology of individuals from both of these populations are known (e.g. interactions with other heavy metals and past and present environmental factors), but it does mean that ecological significance cannot be presumed to follow. What is interesting in this case is that the correlation between metal sensitivity and environment exists in the first place.

4.2.2 Acclimatization and reversible non-genetic differences

If population differences in physiology can be removed by keeping individuals in controlled conditions in the laboratory, those original differences can be attributed to local physiological acclimatization, and in particular, to reversible non-genetic differences. Early work on the ragworm *Hediste* (as *Nereis*) *diversicolor* demonstrated differences in osmoregulation between geographically separate populations inhabiting different salinity regimes (Schlieper 1929). Subsequent studies confirmed such differences and attributed them, in truth without much supporting data, to racial (i.e. genetic) rather than environmental factors. This conclusion found its way into early and well respected physiological textbooks (e.g. Krogh 1939). However, Smith (1955) found that keeping individuals from a number of populations (collected from different parts of the geographical range of the species) under standard laboratory conditions largely removed any differences in chloride ion regulation. So it would appear that what was attributed to physiological differentiation is, in fact, merely a physiological difference due to acclimatization. It would be interesting to know how many of the numerous population differences that have been documented in the past can be accounted for by acclimatization and therefore disappear under controlled laboratory conditions.

Another way of investigating the nature of observed physiological differences is by the use of reciprocal transplant experiments. One of the classic studies is that of Davies (1966, 1967) on respiratory differences between high and low shore populations of the limpet *Patella vulgata*. The weight-specific metabolic rate, measured in the summer, of low shore individuals was greater than that of those from the high shore. Davies (1967) suggested that a reduced metabolic rate in high shore individuals might be advantageous to an animal living in an environment characterized by high temperatures, and where the time available for feeding is reduced. To examine the origin of this difference he transplanted high shore individuals to the low shore and vice versa—controls were obtained by removing individuals from both heights on the shore at the time of transplantation and then replacing them nearby at the same shore height. After 3 months, the transplanted individuals were found to exhibit metabolic rates characteristic of other individuals on the same piece of shore rather than of their source populations (Fig. 4.1). Thus, the difference in metabolic rate is most likely the result of phenotypic responses to the specific environmental factors of high and low shore habitats (see Segal 1956 for a similar situation for two populations of the limpet *Acmaea limatula*).

Unfortunately, not all investigations of this type are quite so straightforward. It has long been known that there are distinct population differences in the physiology of the blue mussel *Mytilus edulis* from the North Sea and from the Baltic (e.g. Schlieper 1957). Furthermore, the populations have been

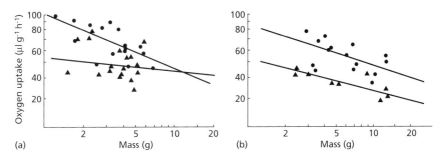

Fig. 4.1 Oxygen uptake of limpets *Patella vulgata* on the shore. (a) Measurements made at 15°C on summer-collected individuals from a high (triangle) and a low (circle) shore population. (b) Measurements at 15°C from high-shore individuals 3 months after transplantation to the low shore (circle) and vice versa (triangle). (After Davies 1967, with permission from Cambridge University Press.)

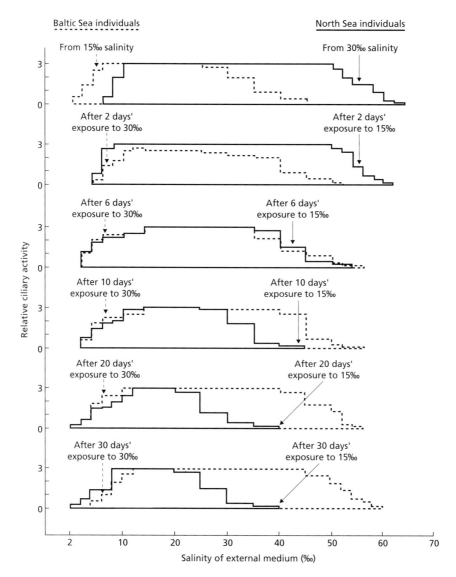

Fig. 4.2 Salinity related changes in relative ciliary activity of the gills of individual mussels *Mytilus edulis* taken from two populations that differ with respect to the salinity regimes they experience in nature. (After Prosser 1973, p. 14.)

shown to be genetically different (Varvio *et al.* 1988; Johannesson *et al.* 1990). And yet such differences in physiology as have been investigated are almost completely reversed as a result of reciprocal transplant experiments (Fig. 4.2; Tedengren *et al.* 1990), supporting Schlieper's view (1960, 1967) that Baltic and North Sea populations should not be regarded as separate physiological races. However, Väinölä and Hvilsom (1991) note that in terms of genetics (i.e. allozyme frequency data) the Baltic population examined by

Schlieper (1960) was most probably 'quite typical North Sea *M. edulis*' but that there was a major genetic transition zone that occurred further south in the Baltic Sea.

Physiological differences between *M. edulis* collected from an estuarine and a fully marine population were also determined largely by environmental rather than genotypic factors (Widdows *et al.* 1984). However, although most of the physiological differences between North Sea and Baltic or estuarine and marine mussels could be accounted for by acclimatization, there did still appear to be a small genetic component in both cases (Widdows *et al.* 1984; Tedengren *et al.* 1990).

Piersma *et al.* (1996) studied differences in body composition and basal metabolic rate in two subspecies of red knot *Calidris canutus*, in the field and under controlled conditions in the laboratory; one that spends the non-breeding season at temperate latitudes, *C. c. islandica*, and one that winters in the tropics, *C. c. canutus*. While there were marked differences in the physiology of field populations, the two subspecies converged to similar body composition and metabolic rates in captivity.

The literature is replete with other examples of local physiological acclimatization, although not uncommonly alongside a genetic component, encompassing a wide range of animal groups, including beetles (Schultz *et al.* 1992), starfish (Prosser 1955), oysters (Stauber 1950; Loosanoff and Nomejko 1957) and fish (Brown and Feldmeth 1971).

4.2.3 Irreversible non-genetic differences

If population differences in physiology cannot be removed by keeping individuals under similar environmental conditions in the laboratory, it does not necessarily follow that this is an example of local genetic adaptation. For example, it is suggested that there were irreversible non-genetic differences in thermal tolerance between Mediterranean and Atlantic populations of the bivalve mollusc *Tellina* sp. (Ansell *et al.* 1986). Similarly, claims were made for effects of temperature on oxygen uptake by clones of the sea anemone *Haliplanella luciae* (Zamar and Mangum 1979). Unfortunately, one often cannot tell from such studies whether the difference is definitely as a result of irreversible non-genetic physiological adjustment. Only if the population differences are removed by breeding both populations through to F_1, or preferably F_2 (the latter is more convincing as it avoids the complication of maternal effects which may still result in differences between the F_1 generations), can one be confident that this is so. Given problems of breeding and culturing animals in the laboratory it is not surprising that few studies have been designed such that a distinction can be made between irreversible non-genetic compensation and genetically fixed physiological differences (Kinne 1962). Studies have examined the effect of an environmental challenge early on in development and how it irreversibly alters structure and function (Sects 2.4.2, 2.4.3), but again breeding studies are rare and so in most cases while physiological adjustment may remain irreversible throughout an individual's lifespan, it is still not clear whether one is dealing with a genetic difference or not.

Physiological heterochrony, which was addressed as a within-individual phenomenon (Sect. 2.4.4), could conceivably be considered at the between-population level as an irreversible event. Here, it is a developmental response to environmental variation within the geographical range of a species, compounded by other environmental factors such as food availability and population density. Physiological heterochrony between different populations may well be a significant source of between-population physiological variation.

The potential for non-genetic differences, reversible or otherwise, is, of course, itself determined by genetic constraints. They consist of responses to, and interactions with, the environment which are themselves genetically determined. Furthermore, the expression of this genetic potential will most likely alter during development (Sect. 2.3). One last note of caution concerns the observation that breeding a population through many generations may itself result in genetic change and an altered physiology. In testing the cold tolerance of five strains (of different origin) of the German cockroach *Blattella germanica*, Haschemi (1992) found that what he termed an outdoor strain (it came from a local refuse tip) was the most tolerant to low temperatures, but its cold-hardiness was dramatically reduced after being kept for 8 years at 22–30°C in the laboratory.

Although irreversible non-genetic local acclimatization may be a widespread phenomenon, there is very little strong empirical evidence that this is so.

4.2.4 Genetic differentiation

If population differences in physiology cannot be removed by keeping individuals (or their progeny, F_1 and F_2) under identical environmental conditions in the laboratory, then it is fairly certain that such differences are due to genetic differentiation. Much of the detailed knowledge of genetically based differences in physiology comes from the fields of toxicology and ecotoxicology. The response of animals to toxic substances in the environment has been relatively well studied, and many resistant populations have been discovered and described. Furthermore, the genetic basis of adaptation of animals to chronic pollution or insecticide deployment has been examined, which throws some interesting light on the nature of genetic differences in physiological traits between populations.

Individuals of the aquatic oligochaete *Limnodrilus hoffmeisteri* collected from a site characterized by chronically high concentrations of the metals cadmium, nickel and cobalt survived longer in metal-rich test solution than did individuals from a nearby uncontaminated site (Fig. 4.3). The resistant populations were often separated from non-resistant populations by a few

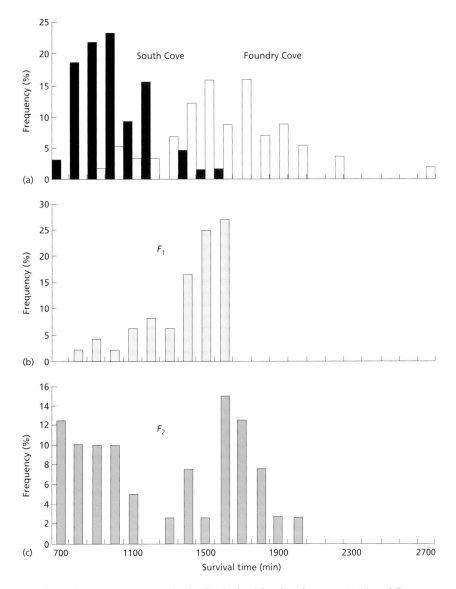

Fig. 4.3 Tolerance to heavy metals of individuals of the oligochaete worm *Limnodrilus hoffmeisteri* from a metal-sensitive (South Cove) and metal-tolerant (Foundry Cove) population: (a) Freshly collected individuals; (b) the result of crossing individuals of both populations (F_1); and (c) their progeny (F_2). (After Martínez and Levington 1996, with permission from the Editor of *Evolution*.)

hundred metres with no apparent barrier between them. When individuals from both populations were crossed, the distribution of survival times for F_1 individuals was left-skewed and the median was within approximately the same range of survival times as exhibited by the resistant population. When taken through to the F_2 generation, the frequency distribution was bimodal and the modes corresponded quite markedly with the parental generation.

The results of this study suggest, although they do not show unequivocally, that a single segregating genetic factor underlies the resistance to heavy metals shown by individuals from the metal contaminated site. The genetics of resistance may be related to evolutionary change in metallothionein genes. Certainly, Maroni *et al.* (1987) found that individuals of the fruitfly *Drosophila melanogaster* with metallothionein gene duplications were more resistant to copper than individuals with only single copies of the gene, although this was not borne out by field studies of populations from polluted and non-polluted sites (Lange *et al.* 1990). Indeed, it is conceivable that differences in resistance could be attributed to the inability of individuals in some populations to synthesize metallothioneins at all. British populations of the bivalve *Macoma balthica* do not seem able to produce metallothioneins, and they are more sensitive to exposure to heavy metals than are introduced populations in California which do seem to have this ability (Rainbow 1997; while there is some doubt as to whether the populations are indeed monospecific, they are certainly very closely related). Genetically determined between-population differences in pollution tolerance have been demonstrated for a number of other animal groups, including springtails (Posthuma *et al.* 1993), slugs (Grenville and Morgan 1991) and the ragworm *Hediste* (as *Nereis*) *diversicolor* (Bryan and Hummerstone 1971) (see Maltby (in press) for a list and details of metal-tolerant invertebrate populations), and there is a respectable literature on the differential response of different populations to pollutants (cf. Maltby *et al.* 1987; Nelson 1990; Chan *et al.* 1992; Nelson and Mitchell 1992; Spicer *et al.* 1998).

Comparing the percentage mortality of the mosquito *Aedes aegypti* for different doses of an insecticide permethrin, there was a clear difference between sensitive (S) and resistant (R) strains, as might be expected (Fig. 4.4). When they were crossed, the next generation (F_1) showed mortality intermediate to that of the S and R strains. Furthermore, backcrosses (i.e. $F_1 \times$S and $F_1 \times$R) produced lines which represent a 50:50 mix of F_1-like and S (or R)-like types. The shape of the mortality curves for the backcross throws some light on the

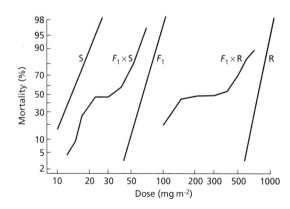

Fig. 4.4 Tolerance to the insecticide permethrin of sensitive (S) and resistant (R) strains of the mosquito *Aedes aegypti*, their crosses (F_1) and backcrosses ($F_1 \times$S, $F_1 \times$R). (After Mallet 1989.)

mechanisms underlying the difference between the strains. It provides good evidence that the population difference observed was due to the actions of a single gene (or perhaps at most several tightly linked genes), as the associated action of a number of genes, each producing a small effect (polygenic inheritance), would be expected to produce a straight trajectory on the figure in the backcrosses (Fig. 4.4). This example also illustrates the importance of the environmental variable of interest itself—in this case, concentration of insecticide in the environment—in selection. At low doses (e.g. $30\,mg\,m^{-2}$) homozygous S genotypes were eliminated and resistance was dominant, but as the dosage was increased (e.g. to $300\,mg\,m^{-2}$) heterozygotes for resistance were also affected, and resistance became recessive (Mallet 1989).

The existence of genetically determined population differences in physiological traits can have serious implications for some of the analyses carried out in the developing field of ecotoxicology. For example, many of the lists of pollution-sensitive or pollution-tolerant species assume a fixed species physiology, and that this physiology accurately reflects that of the species when it is present in any given community. However, there are cases where species thought of as pollution-sensitive include some populations that are clearly not (e.g. Spicer *et al.* 1998). The extent to which this is a significant, or even common, problem has yet to be established. Nevertheless, until more is known of population differences in the physiologies of at least some of the key species employed in monitoring programmes, one should err on the side of caution.

It is often assumed that an effective way of substantially reducing the physiological variation encountered in experiments is to examine genetically identical animals or clones. This, for example, is the basis of using inbred rat strains (e.g. Wistar strain) or clonal and asexually reproducing animals, such as the waterflea *Daphnia*, in toxicological tests. Standardization both within and between testing laboratories is understandably a high priority for governmental agencies, particularly when the results of such tests are to be used for legislative purposes (Calow 1996). And yet such standardization has sometimes proved elusive even using clones (Sect. 3.5.1). Somewhat disconcertingly, a recent study shows that clonal populations of hydrobiid snails are as phenotypically variable as are sexually reproducing populations (Forbes *et al.* 1995).

As well as evidence from pollution studies, there have also been a number of more physiological investigations, where breeding experiments have convincingly shown that population differences have a genetic basis, even if the exact nature of that basis is not always clear. Because of the interest in their aquaculture, various commercially important fish species have received particular attention. There are population differences in thermal tolerances of juvenile large mouth bass *Micropterus salmoides salmoides* ('northern') and *M. s. floridanus* ('Florida'), which persist into the F_2 generation, with hybrids of individuals from the two populations showing intermediate values (Fields *et al.*

1987). The authors of this study found no correlation between thermal tolerance and genotype (as estimated from measurement at three enzyme loci). Genetic differences in thermal tolerance have also been demonstrated between two strains (Auburn University, Egypt and Auburn University, Ivory Coast) of the fish *Oreochromis niloticus*, by taking them through to F_1, F_2, and backcross hybrids (Tave *et al.* 1989), and between two populations of eastern mosquitofish *Gambusia holbrooki* in South Carolina (USA), one in an ambient temperature pond and the other in a nearby pond heated to near-lethal temperatures by nuclear reactor effluents for 60–90 mosquitofish generations (Meffe *et al.* 1995).

Again, primarily due to their commercial value, the physiology of different strains of brine shrimp *Artemia* has been compared. It seems that some populations may be better genetically adapted than others to cope with environmental challenge. For example, the effect of temperature on metabolism is smaller (Varó *et al.* 1991) and tolerance of radiation is greater (Metalli and Ballardin 1972) in polyploid when compared with diploid populations, and there are differences in thermal tolerance between bisexual and parthenogenetic *Artemia* strains (Amat 1983; Vanhaecke *et al.* 1984). Genetically determined physiological differentiation has also been demonstrated for salinity tolerance and osmoregulation in different populations of the copepod *Tisbe furcata* (Battaglia and Bryan 1964; Battaglia 1967) and the crab *Sesarma reticulatum* (Staton and Felder 1992), and for salinity tolerance of different populations of the dogwhelk *Nucella lapillus* (Kirby *et al.* 1994), sea urchins (Stickle *et al.* 1990) and ascidean sea squirts *Ciona intestinalis* (Dybern 1967). Other examples of genetically determined population differences include melting points of cuticular lipids in grasshoppers (Gibbs and Mousseau 1994), energy budgets in scallops (Volckaert and Zouros 1989) and metabolism in snakes (Beaupre 1993). Even estuarine species with larval dispersal can exhibit marked physiological differentiation as well as local acclimatization (Stauber 1950; Loosanoff and Nomejko 1957; although cf. Sect. 4.3.2). Interestingly, although there was no difference in exercise performance between an open water (Atlantic) and a brackish lake (Nova Scotia, Canada) population of Atlantic cod *Gadus morhua*, the physiological mechanisms underpinning performance (metabolic, ventilatory and cardiac) were quite different in each case (Nelson *et al.* 1994).

Notwithstanding earlier comments on whether or not there was a genetic basis to the physiological differentiation between North Sea and Baltic mussels *Mytilus edulis* (Sect. 4.2.2), there is good evidence that such differentiation occurs in this species at other locations with respect to salinity. Clinal variations in nitrogen metabolism are correlated with salinity in *M. edulis*. Leucine aminopeptidase (*Lap*), which removes *N*-terminal amino acids from polypeptides, occurs in three alleles, each named from electrophoretic mobility (98, 96, 94 — see review by Koehn 1983). One allotype (*Lap*-94) occurs with a frequency of 0.55 in specimens from ocean water at the tip of Long Island (USA), and at a frequency of 0.15 in specimens from the dilute water on inner Long

Island Sound (Hilbish 1985). Specimens with *Lap*-94 accumulate more free amino acids than specimens with other allotypes. In a dilute medium such accumulation is disadvantageous, in that excretion of amines and of some ammonia is involved in cell volume regulation. Juveniles show high mortality of *Lap*-94 genotypes in dilute sea water, hence the clinal distribution of the three allotypes results from differential mortality during migration from the ocean to inner Long Island Sound (Hilbish *et al.* 1982, 1994; see also Deaton *et al.* 1984; Hilbish and Koehn 1985a,b).

The examples given above on pollution resistance or resistance to insecticides seem to indicate that the mechanisms underlying physiological differentiation may be very simple, linked to the actions of a single gene (or tightly grouped gene complex) and may even involve gene duplication. The extent to which this is the case for population differentiation of other physiological traits remains to be seen. What is clear is that while it may be possible to correlate physiological and genetic diversity when the physiological trait under examination is somehow related to the specific genetic material under scrutiny, there seems little reason to believe that such a correlation will exist between variation in a physiological trait and genetic variation more generally.

Genetically determined physiological differentiation of east and west Atlantic populations of the barnacle *Semibalanus* (as *Balanus*) *balanoides* was elegantly demonstrated by Crisp (1964c) using transplant experiments. However, here and in many of the studies considered up until this point, the wrong impression may be given by considering environment and genetic explanations of the origins of population differences to be mutually exclusive. Often both may be involved. Pierce and Crawford (1996) systematically determined the prevalence of altered enzyme expression by examining all the enzymes in a single key metabolic pathway, namely glycolysis in heart tissue of the killifish *Fundulus heteroclitus*. Both variation between populations and the effect on enzyme expression of being kept at different temperatures were investigated. Of the 10 cardiac glycolytic enzymes detected, two (phosphoglucoisomerase and aldolase) were expressed at a significantly greater level in a northern population than in a southern one. Neither was affected by the temperature at which individuals were kept. The expression of two other enzymes (phosphoglyceromutase and enolase), although they were for the most part similar in both populations, was significantly greater as a result of being maintained under cold conditions. These findings indicate that both evolved differences between populations, and physiological adjustment within a population occur in a single pathway and can potentially compensate for the differences in the thermal environment. Furthermore, all four enzymes that show variation are equilibrium enzymes that are traditionally considered incapable of producing compensatory metabolic changes. Consequently, these results call into question the assumption that only variation in non-equilibrium enzymes is important for the regulation of metabolism.

That such genetic differentiation between populations of a species seems common, means that great care must be exercised when collecting individuals for use in laboratory experiments, and in particular when those experiments are aimed at making species comparisons. At the very least, the source of the population should be acknowledged, although for some objectives, nothing less than knowing the breadth of physiological variation among populations of a given species will suffice.

4.2.5 Demographic differences

Typically, as for within-population variation (Sect. 3.1), laboratory studies of between-population physiological differences, regardless of whether such differences are genetically or environmentally determined, control for the effects of body size variation. And yet rarely are different populations characterized by similar demographic features (e.g. body sizes, ages and developmental stages); these features are themselves products of the interaction between prevalent environmental conditions and genetics. As we have seen, it is not always safe to assume that in terms of their physiologies, small individuals are large individuals writ small (see Sect. 2.3.3). Unless there is strong evidence to the contrary, so doing serves to obscure another important and ecologically relevant potential source of between-population physiological variation. For example, the maximum thermal tolerances (CT_{max}) for two populations of the widespread beachflea *Orchestia gammarellus*, one from northern Scotland and the other from south-east England, have been examined (Gaston and Spicer 1998b). Controlling for body size tends to underestimate the difference in CT_{max} between individuals from the two populations. This is because animal size influences CT_{max} and the southern population is comprised of larger individuals, giving rise to markedly different frequency distributions of tolerances, which could not be accounted for by acclimation (Fig. 4.5).

4.3 Types of population similarity

As mentioned at the outset (Sect. 4.1), there are examples of cases in which individuals from different populations do not appear to exhibit differences in particular physiological traits. There are several reasons why this might occur.

4.3.1 Similarity in environmental conditions and absence of capacity for local acclimatization

If most of the potential between-population variation in a particular physiological trait is as a result of acclimatization, then one would expect that populations inhabiting locations with very similar (key?) environmental conditions would display little variation in that physiological trait. There is no difference

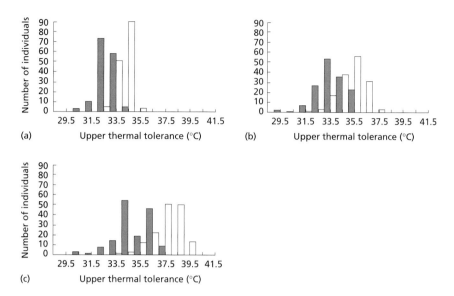

Fig. 4.5 Frequency distributions of upper thermal tolerance of individuals of the beachflea *Orchestia gammarellus* for a population from northern Scotland (shaded bars) and one from south-east England (open bars). Individuals kept at (a) 5°C, (b) 10°C and (c) 15°C. (After Gaston and Spicer 1998b.)

in the effect of temperature on aerobic metabolism for four populations of diploid brine shrimp *Artemia*; two Spanish populations (La Mata and Cádiz), one from the Canary Islands (Lanzarote) and one from the Great Salt Lake (Utah, USA) (Varó *et al.* 1991). These crustaceans inhabit the same kind of environment (ephemeral ponds subject to pronounced fluctuations in environmental factors) at each of these locations. Likewise, Schultz *et al.* (1992) found that the mean body temperatures and water-loss rates of foraging adult tiger beetles *Cicindela longilabris* did not vary significantly among four disjunct populations (Arizona, Colorado, Maine, and Wisconsin (USA)). Initial differences in metabolic rate disappeared after individuals from different populations were kept in the laboratory for a period. The absence of population differences was attributed to the fact that this species occupies climatic refugia at lower latitudes.

The sea urchin *Stronglyocentrotus purpuratus* has a range extending from Alaska (USA) down into Mexico (Ricketts and Calvin 1968). However, in common with many intertidal animals, it is not found at the same height on the shore throughout this range; individuals occur intertidally at high latitudes and subtidally at low latitudes. This probably serves to reduce differences in the key environmental conditions (i.e. environmental temperature variations, but notice, not tidal inundation) experienced by the different populations. Individuals of different geographical origins have a set upper thermal limit of about 23.5°C (Farmanfarmaian and Giese 1963). Either the variation in whatever are the key environmental factors is similar for each of the locations and/or what is seen reflects the fact that this species has no capacity for local acclimatization. More generally, terrestrial animals are known to occupy different altitudes at different latitudes, which again serves to ameliorate the environmental variation they would otherwise experience.

4.3.2 Gene flow

Between-population differences in physiology would not be expected, other than perhaps by local physiological acclimatization, if there is significant gene flow between the populations. Bertness and Gaines (1993) provide a nice example of this. Thermal stress is a strong selective force on the barnacle *Semibalanus balanoides* in estuarine habitats. They found that in an estuary with a short flushing time recruits do not show local acclimatization to thermal stress because flushing likely mixes larvae from a variety of sources. In contrast, in an estuary with a slow average flushing time, when this rate is realized, recruits are found to show local acclimatization to thermal stress and when the rate is not realized (as a result of heavy rains) they are not. The slow flushing time prevents the mixing of larvae, resulting in recruits being the progeny of individuals who have themselves survived the local thermal stress regime.

Whether or not there is significant gene flow between populations may be particularly important for the type of physiologies exhibited by individuals from populations at the edges of the range of a species (Sect. 4.5.2). This should prove a fruitful area of study.

4.4 Spatial patterns in between-population variation

In many instances, where they are present, differences in physiological traits between populations can be systematic across space, constituting broad geographical patterns of physiological variation. In particular, one might expect patterns to exist with regard to three major gradients, respectively, in latitude, altitude and depth. Changes in all of these are associated with changes in multiple environmental parameters. We shall examine examples of each of these clines in turn.

4.4.1 Latitude

With an increase in latitude there is, among other things, a decrease in mean environmental temperature. Generally speaking, variations in temperature extremes also tend to increase, except in the case of polar marine systems where water temperatures show little variation. It would be surprising, therefore, if between-population variation in physiological traits was not often associated with latitude. Here we consider two traits which have perhaps received most attention in this regard, thermal tolerance and metabolic rate.

Thermal tolerance

As early as the beginning of this century, it was shown that the jellyfish *Aurelia aurita*, which literally has a worldwide marine distribution, displays a decrease in upper critical temperature with an increase in latitude (Mayer 1914). The horseshoe crab *Limulus polyphemus* also exhibits such a latitudinal change in upper critical temperature (Mayer 1914).

In fact, relatively few studies have examined clinal physiological variations with latitude. Most compare the physiologies of two latitudinally separate populations. The fiddler crab *Uca rapax* lives in the supratidal zone throughout its geographical range (USA to Brazil). Examination of data on upper thermal tolerances of two geographically separated populations (Cuba and Brazil; data from Vernberg and Vernberg 1972, Fig. 2, p. 68) clearly reveals that individuals from populations inhabiting lower latitudes show a greater tolerance of high temperatures than do individuals from populations inhabiting higher latitudes ($LD_{50} = 40$ and 17 min, respectively, all 'warm acclimated' individuals tested at 42°C). Similarly, laboratory cultures of the crinoid *Antedon petasus* from the Norwegian coast had an upper thermal tolerance of 14°C, which was 20°C less than measured for individuals of populations from Tobago (Hyman 1955). A non-reversible decrease in upper thermal tolerance with increasing latitude has also been demonstrated for bivalve molluscs (Ansell *et al.* 1986) and beachfleas (Gaston and Spicer 1998) (cf. Sect. 5.4.1 species vs. latitude). Numerous other between-population studies could be cited (e.g. juvenile red drum (fish) *Sciaenops ocellatus*—Ward *et al.* (1993); hatchling painted turtles *Chrysemys picta*—Packard and Janzen 1996) that have confirmed this general pattern of increasing upper and lower thermal tolerances with decreasing latitude, although there are some exceptions (e.g. musk turtles *Sternotherus odouratus*—Ultsch and Cochran 1994). Unfortunately, for many of the between-population studies, the extent to which they concern reversible acclimatization is not clear, but the potential for generating patterns of a non-reversible genetic character lies in strong latitudinal clines in climatic (particularly temperature) and other environmental variables.

Indeed, non-reversible latitudinal clines in physiological variation plainly do occur. In a study of the house spider *Achaearanea tepidariorum* from six latitudinally separate populations (26°12′, 31°34′, 33°34′, 35°41′, 39°41′ and 43°03′N in Japan, where the proportion of cold winters over the past 43 years has been 0, 2.3, 2.3, 9.3, 100 and 100%, respectively), Tanaka (1996) found that while there was no difference in the supercooling point of summer-collected individuals, in the case of winter-collected animals, it was lower at higher latitudes (Fig. 4.6). Examination of the frequency distribution plots of supercooling ability during the winter for each of the six populations indicates

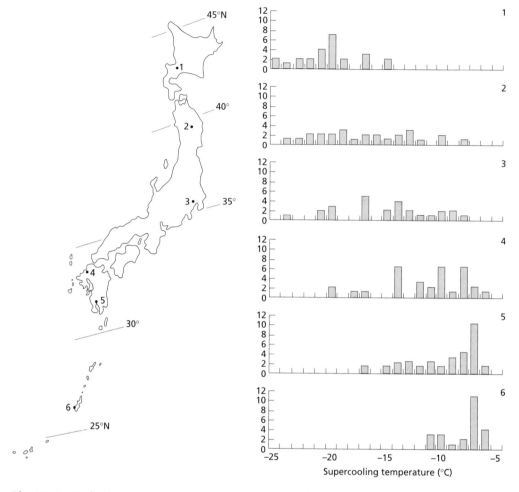

Fig. 4.6 Geographical occurrence of populations of the house spider *Achaearanea tepidariorum* in Japan and the frequency distributions of supercooling temperatures of winter-collected individuals from each of the populations. (After Tanaka 1996.)

that the latitudinal cline observed is achieved from one latitudinal location to the next by a proportional increase in the number of individuals with lowered supercooling ability and a greater degree of supercooling ability at the higher latitudes. Tanaka (1996) notes that from north to south the percentage of winter-collected individuals that maintain the supercooling ability measured in summer-collected individuals is 100, 75, 58.1, 23.1, 11.5 and 0%. He also shows that these differences have a genetic basis, and suggests that while northern individuals lower their supercooling ability in winter and southern individuals do not, the coexistence of spiders with both low and high supercooling ability in the warm–temperate zone is probably related to the unpredictability of annual variation in winter climate. If correct, this provides a good example of increased physiological variation accompanying increased environmental variation, where the most physiological variation is encountered in the middle of the latitudinal cline.

In this connection, a study of three populations of the bluegill *Lepomis macrochirus* inhabiting different thermal regimes is of interest (Holland *et al.* 1974). Individuals from the heated end of a large reservoir potentially encounter a greater range of environmental temperatures than either those living in a small isolated pond which directly receives heated effluent or those from the far end of the reservoir which is unaffected by the effluent (Fig. 4.7). Consequently, individuals from the heated end exhibit a much wider range of thermal tolerances; the frequency distribution is possibly right-skewed (Fig. 4.7b), indicating that while a large proportion of the data are in the same range as equivalent data from individuals exposed to ambient temperatures (Fig. 4.7c), there are some heat tolerant individuals which share a similar range of tolerances to individuals from the small hot pond (Fig. 4.7a). Thus, although not over a latitudinal cline, this provides another example of increased environmental variability, in this case temperature, being accompanied by greater variation in a related physiological trait, thermal tolerance.

Aerobic metabolism

The effect of latitude on differences in the respiratory physiology of populations belonging to the same species has received some attention. Latitudinal divergence in aerobic metabolism between different (again usually only two) populations has been documented, for example, for some fiddler crab *Uca* species (Démeusy 1957; Vernberg 1959), for the high shore crab *Pachygrapsus crassipes* (Roberts 1957), two malanid polychaete species *Clymenella torquata* and *C. mucosa* (Mangum 1963), oysters *Crassostrea virginica* (Dittman 1997), the fruitfly *Drosophila melanogaster* (Berrigan and Partridge 1997; Berrigan 1997), the tiger beetle *Cicindela longilabris* (Schultz *et al.* 1992) and the lizard *Sceloporus occidentalis* (Tsuji 1988) (cf. also rate of water propulsion and heart rate of mussels as a function of latitude; Rao 1953; Pickens 1965). Generally speaking, individuals from the more northerly population have elevated rates of aerobic metabolism (when compared at the same temperature). However, there are a few studies which show no clear differences, for example for North temperate and sub-Antarctic polychaete worms *Thelepus setosus* (Duchêne 1985), the beetle *Calathus melanocephalus* from northern Norway and the Netherlands (Nylund 1991) and the fiddler crab *Uca rapax* (Vernberg and Vernberg 1996a,b). Apart from the fact that they derive predominantly from studies comparing two sites, the data available for studying latitudinal effects in some cases are also open to the criticism that acclimation treatments are not included in the experimental design, although there are exceptions (e.g. Mangum 1963; Vernberg and Vernberg 1996a,b; Tsuji 1988).

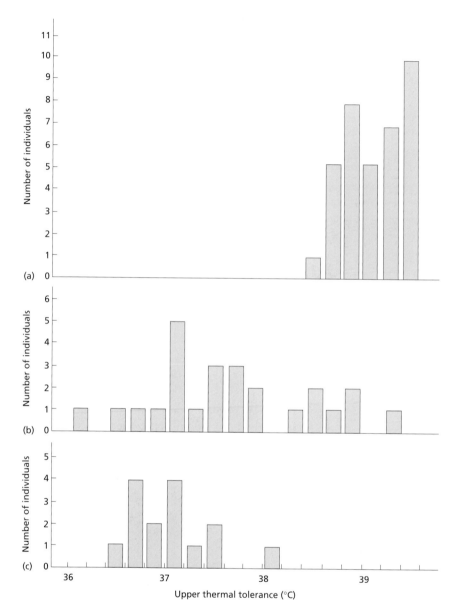

Fig. 4.7 Upper lethal temperatures of bluegill *Lepomis macrochirus* from (a) a small enclosed pond receiving heated effluent from a nuclear reactor ($n=36$); (b) the warm end of a large reservoir also supplied with heated effluent (1100 hectares, the heated end of the reservoir ranges between 30 and 40°C during the summer and 20–30°C in winter, although there is a dramatic reduction to ambient temperature within 100 m of where the effluent enters the reservoir) ($n=25$); and (c) a section of the reservoir unaffected by the effluent and therefore subject to ambient temperature conditions ($n=15$). (After Holland *et al*. 1974, with permission from The University of Chicago Press.)

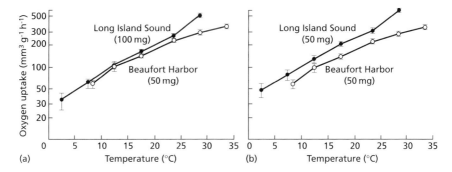

Fig. 4.8 (a) Actual and (b) size-corrected rates of oxygen uptake of individuals of the polychaete worm *Clymenella torquata* from two different populations. Values are means ± 1 SD. (After Mangum 1963, with permission from Elsevier Science.)

Mangum (1963) measured oxygen uptake as a measure of metabolism (for individuals kept at a range of different temperatures) of two latitudinally separate populations from each of two species of polychaete, *Clymenella torquata* (41°15′N, 34°40′N) and *C. mucosa* (34°40′N, 18°N). She found that, not correcting for differences in body size between the populations (the mean body size of the more northerly population is about half that of the southerly one), the rates of metabolism of *C. torquata* at different temperatures were very close, although they were still significantly different for test temperatures of, or greater than, 17.5°C. Individuals from the more northerly population were characterized by a metabolic rate that, where differences were significant, was consistently higher than that of the more southerly one (Fig. 4.8a). When the data were standardized with respect to body weight, this difference was accentuated across the full range of temperatures investigated (Fig. 4.8b). The difference is most likely to be as a result of physiological adjustment rather than selection, as mortality at each experimental temperature was negligible. In the closely related *C. mucosa*, the pattern seemed to be reversed (warmer water individuals had a consistently greater metabolism than cold-water individuals), although standardizing for body weight resulted in a reversal of this pattern (in common with *C. torquata*, the cold-water population has a consistently greater metabolism than the warm-water population). Given that it could not be accounted for by keeping individuals at the same temperature, the difference is probably genetic, although irreversible non-genetic acclimatization cannot be ruled out.

Vernberg and Costlow (1966) have addressed this last problem directly in a study of oxygen uptake of individuals from three latitudinally separate populations of the fiddler crab *Uca pugilator*. Here individuals from each population were hatched and reared under identical laboratory conditions. While this does not rule out the possibility of maternal effects, it does bring one a step closer to attributing differences (if observed) to genotypic rather than pheno-

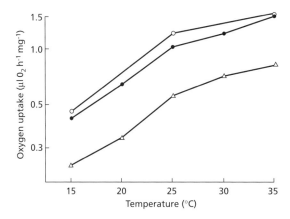

Fig. 4.9 Rates of oxygen uptake of individuals of the fiddler crab *Uca pugilator* reared in the laboratory, and originating from populations at three latitudes, from north to south: Massachusetts (open circles), North Carolina (closed circles) and Florida (open triangles), USA. (After Vernberg and Costlow 1966, with permission from The University of Chicago Press.)

typic variation. They found that oxygen uptake was greater in 'cold'-water populations (North Carolina and Massachusetts, USA) when compared with 'warm'-water populations (Florida, USA) at each of the temperatures investigated (Fig. 4.9), which corroborates the findings of Mangum (1963). Indeed, despite some exceptions, it seems to be a general pattern that individuals from populations at high latitudes have a greater metabolism than those from populations at low latitudes, when both are standardized for body size and tested under identical laboratory conditions. As body size within a species seems commonly to decrease with a decrease in latitude, this will result in an amelioration of the decline in metabolism towards lower latitudes if actual metabolic rates (uncorrected for body size) are compared. For the moment we will be content to document these general patterns, leaving the investigation of their significance to the next chapter when between-species latitudinal differences in physiology will be considered (Sect. 5.4.1).

Enzymes are often the tools employed in implementing physiological regulations and acclimatization responses and so clinal variation in allozyme frequencies is potentially of direct relevance to the characterization of between-population physiological variation. However, without more specific information on the physiological role of these enzymes or what the variation detected covaries with, it is perhaps unwise to draw any firm conclusions on clinal variations in physiology from such studies. More (obviously) 'physiologically relevant' allozyme studies are required for this.

Two examples of allozyme clines for a key enzyme involved in anaerobic metabolism of vertebrates and crustaceans, and thus possibly of direct physiological interest, are as follows. First, the relative proportions of two alleles for lactate dehydrogenase (LDH), termed A and A', in the crested blenny *Anoplarchus perpurescens*, vary along a latitudinal cline on the Canadian west coast (Johnson 1971). The frequency of A' clearly increases with a decrease

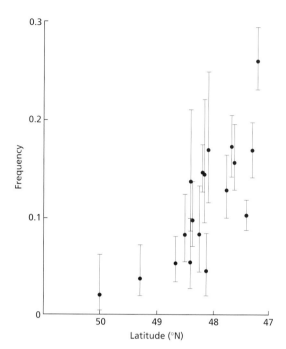

Fig. 4.10 Frequency of A′ allele for the enzyme lactate dehydrogenase in individuals of the crested blenny *Anoplarchus perpurescens* from 28 populations distributed along a latitudinal cline on the west coast of Canada. Each point represents a mean value and 90% confidence intervals. (After Johnson 1971.)

in latitude (Fig. 4.10). Furthermore, this pattern can be produced in the laboratory by rearing fish at different temperatures. There is therefore a genetic basis for clinal variation in phenotypic response. However, it is still one step removed from the possible ecological relevance of this cline.

Second, in the killifish *Fundulus heteroclitus*, a number of different isozymes, including LDH, show clinal variation (Fig. 4.11). Although the physiological significance of all of these allozyme patterns is not always clear, Prosser (1986) felt in such cases that the differences observed were sufficient to propose the existence of subspecies, and suggested environmental temperature as the selective force involved. In the case of LDH, there are two alleles (in this study termed Bᵃ and Bᵇ) which vary with latitude. Maximum swimming speed is greater in populations with higher proportions of LDH-Bᵇ than LDH-Bᵃ and the concentration of adenosine triphosphate (ATP) in red blood cells is proportional to the swimming performance (the ratio of the oxygen-binding pigment haemoglobin to ATP within red blood cells alters the oxygen-binding properties of the pigment). At the northern edge of the range of *F. heteroclitus* (Halifax, Canada), the fish are homozygous for LDH-Bᵇ and at the southern end (Florida, USA) are homozygous for LDH-Bᵃ; the two genotypes occur in latitudinally related proportions throughout the latitudinal extent of this species (Place and Powers 1979; Di Michele and Powers 1982; Di Michele *et al.* 1986; Powers *et al.* 1986).

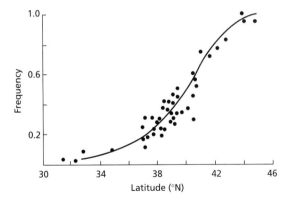

Fig. 4.11 Frequency of B^b allele for the enzyme lactate dehydrogenase in individuals of the killifish *Fundulus heteroclitus* from populations distributed along a latitudinal cline on the east coast of the United States. (After Clarke 1993a; based on data from Powers *et al.* 1986.)

4.4.2 Altitude

With an increase in altitude there is a decrease in air density and oxygen partial pressure (and often humidity) and an increase in exposure to solar energy during the day and in rate of loss of longwave radiation at night. Consequently, animal populations existing at high altitudes are exposed to, among other things, wider temperature fluctuations and a marked hypoxia. These changes occur over much shorter absolute distances than do latitudinal changes in environmental conditions—a change of 2–3°C is experienced over some 10° of latitude or approximately 700 m of altitude in the hills of northern Britain (Whittaker and Tribe 1996). Just as populations at different latitudes exhibit variation in physiological traits, one might therefore expect populations at different altitudes to do so.

A variety of examples demonstrates that this is indeed correct. Miller and Packard (1977) studied the upper thermal tolerance (CT_{max}) of 14 populations of the chorus frog *Pseudacris triseriata* from different altitudes (1530–3036 m elevation, north-central Colorado, USA). They controlled for sex (only males were used) and body size (as body size decreases with increasing altitude, the investigators examined the CT_{max} of sets of individuals from one of the populations that differed in their body size—they found no effect of body size on CT_{max} in this experiment, although they still include it as a factor in their analysis of variance along with time in the laboratory). Individuals were kept for three to four weeks at either 5 or 20°C and then the CT_{max} of each was determined. There were significant inverse relationships between CT_{max} and altitude for both of the experimental temperatures (Fig. 4.12). The effect of altitude accounted for 94% and 82% of the variation at the two temperatures. In sum, there was a decrease in resistance to high temperatures with an increase in altitude (most of the variance is accounted for by altitudinal differences) which is likely to have a genetic basis, although irreversible non-genetic acclimatization again cannot be ruled out.

Upper and lower thermal tolerances were measured for individuals from

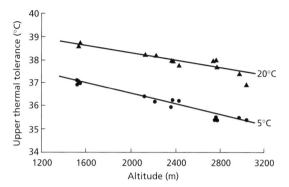

Fig. 4.12 Upper thermal tolerance (CT_{max}) for chorus frogs *Pseudacris triseriata* collected at a number of different altitudes and then kept at either 5°C (circles) or 20°C (triangles) in the laboratory. (After Miller and Packard 1977, with permission from The University of Chicago Press.)

six populations of the gecko *Tarentola boettgeri* collected at different altitudes on Gran Canaria (Portugal) (Brown 1996). In freshly collected individuals both upper and lower thermal tolerances decrease with increasing elevation. However, after 11 weeks of being kept in the laboratory this pattern disappears for upper, but not for lower, thermal tolerance. Thus, there is an increase in the range of thermal tolerance with increasing altitude in this species (see Sect. 5.4.2).

Elevation of rates of oxygen uptake with increase in altitude at the population level has been demonstrated for a number of invertebrate (e.g. insects—Kennington 1957; Massion 1983; Hadley and Massion 1985; Chown *et al.* 1997) and vertebrate groups. Hayes (1989) found that field metabolic rates of deer mice *Peromyscus maniculatus* are greater (57%) at high altitude (3800 m) than at low altitude (1230 m). However, maximal rates of metabolism are independent of altitude. Consequently, during activity, individuals at high altitude operate closer to the maximal sustainable rate of metabolism than do individuals at low altitude. The oxygen binding properties of blood from individuals belonging to 10 different populations of this same species have been studied over an altitudinal cline (0–4 km) by Snyder and coworkers (Snyder 1982; Snyder *et al.* 1982). They find that there is an increase in haemoglobin oxygen affinity with increasing altitude. While such a feature seems clearly adaptive, as highlighted by Burggren *et al.* (1991, p. 477) there are not inconsiderable difficulties in correctly interpreting such patterns.

Two populations of the rodent *Spalacopus cyanus*, one from high altitude and cold burrows in the Andean mountains and the other from low altitude and warmer burrows near the Pacific ocean in central Chile, show similar thermoregulatory ability over a wide range of environmental temperatures (Contreras 1986). High-altitude individuals are characterized by a lower mass specific basal metabolic rate than low-altitude individuals, although the former have a larger body size (mass) than the latter.

Hayes and Shonkwiler (1996) investigated the effects of large altitudinal differences (approximately 3 km) on water fluxes of free-living deer mice. In absolute terms, high-altitude mice had higher water fluxes and field metabolic rates, but were smaller, than low-altitude mice. Hence, altitude had substantial overall effects on water fluxes, but these effects are likely attributable to altitudinal differences in the thermal environment, not to changes in the partial pressure of oxygen.

For the few animal species examined, populations that are found at high altitudes have a greater range of thermal tolerance and higher metabolic rates than populations found at lower altitudes.

4.4.3 Depth

The deep sea is the largest single environmental zone in the world, comprising 70% of the earth's surface, with 76% of that area in excess of depths of 4 km (Meadows and Campbell 1988, p. 7). A number of inland lakes also achieve great depths (e.g. Lake Baikal). With an increase in depth there is an increase in hydrostatic pressure, a decrease in light intensity and a decrease in temperature in both tropical and temperate regions. Of each of the three spatial clines, depth has perhaps been studied the least at the between-population level. The few experiments reported have examined physiological responses to pressure of essentially shallow water species. For example, there are differences in the effect of pressure on ciliary activity on the gills of mussels *Mytilus edulis* between populations collected from the North Sea and from the Baltic (Ponat 1967). However, this difference was shown to be a phenotypic response to salinity (interacting with pressure) rather than an effect of pressure *per se*.

The paucity of studies is not because of a lack of markedly eurybathic species. Examples include the worm *Pomatoceros triqueter* (0–3 km), the polychaete *Amphicteis gunneri* (0–5 km) and the pogonophoran worm *Siboglinum caulleryi* (0–8 km) (Tait and Dipper 1998, p. 126). Rather, it is, as far as we are aware, simply that no one has attempted to compare the physiology of populations of such species. However, given both the logistic difficulties and the financial implications of carrying out such a study, that the comparison has not been made is unsurprising.

4.5 Geographic ranges

4.5.1 Climate and occurrence

The existence of between-population differences in physiological traits, and of systematic spatial patterns in these differences, is strongly suggestive of a general role of physiology in determining where a species can and cannot occur (see Chown and Gaston 1999). Of course, in the extreme this must be true, a species cannot occur (or, at least, cannot persist other than by repeated invasion) in areas with environmental limits which lie beyond the physiological tolerances and capacities of its individuals. Of more interest, and echoing previous discussion about between-individual variation (Sect. 3.4), is whether particular species do not occur in areas which lie within the bounds of their tolerances and capacities. The huge numbers of animal species which have established successfully in areas in which they previously did not occur but to which they have accidentally or intentionally been introduced by human activities shows that this is so (for reviews see Elton 1958; Ebenhard 1988; Drake *et al.* 1989; Hengeveld 1989; Case 1996; Williamson 1996). The set of climatic conditions under which a species is observed to occur, its climate envelope, has often been used to predict the distribution it will attain in an area to which it has been, or may be, introduced (climate matching). Such an approach has been used in the past to predict the likely spread and distribution in the USA of a number of introduced bird (Tworney 1936) and pest species (Cook 1924; Pepper 1939) and, more recently, the spread and distribution in Australia of the introduced European or German wasp *Vespula germanica* (Spradbery and Maywald 1992). Similarly, Sutherst *et al.* (1995) predict the potential geographical distribution of the cane toad *Bufo marinus* in Australia on the basis of climatic parameters which were found best to describe its native distribution (north-western Mexico to southern Brazil) (Fig. 4.13).

Of course, in most cases, these introductions concern the movement of individuals between continents or biogeographic regions, across distances and other physical barriers (e.g. water bodies, land masses, mountain chains and ocean currents) which natural dispersal would be exceedingly unlikely to achieve (at least in sufficient numbers to enable populations to establish). The question thus remains as to whether on a less dramatic scale species do not occur in areas which lie within their physiological tolerances and capacities. Explicit attempts to test this experimentally for animal species are remarkably scarce (in contrast, there have been a number of such studies for plants). None the less, several lines of evidence suggest that it must be so. These include (i) the observation that the absence of some species from otherwise apparently suitable sites is a function of the degree of isolation of those sites (e.g. Gilbert 1980; Harrison *et al.* 1988; Lawton and Woodroffe 1991; Micol *et al.* 1994); (ii) the success of some programmes carried out for the purposes of conservation which have translocated individuals into areas in which previously they did not occur (e.g. Griffith *et al.* 1989; Wolf *et al.* 1996; Lomolino and Channell 1998); (iii) the recognition that some species exist as metapopulations, in which populations at individual sites become extinct and are recolonized as a result, at least in part, of stochastic processes (e.g. Levins 1969; Hanski 1982; Hanski and Gilpin 1997); and (iv) the finding that the ranges of some species expand when essential resources (e.g. hosts) are provided in previously unoc-cupied areas or competitors or predators are removed from them, or that their ranges contract when essential resources are lost or competitors or predators are added (e.g. Lawton and Woodroffe 1991; Brönmark and Edenhamn 1994). The reduction in the geographical ranges of many animal species as a direct or indirect consequence of human activities provides ample examples of

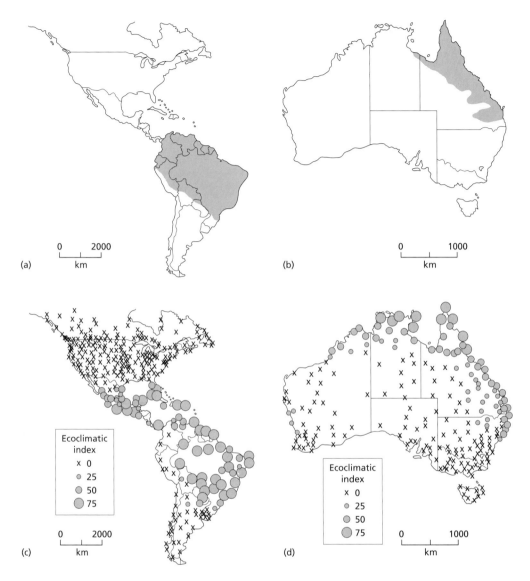

Fig. 4.13 Known distribution of the cane toad *Bufo marinus* in (a) the Americas and (b) Australia, and the distribution predicted using CLIMEX for the two regions (c,d). (After Sutherst *et al.* 1996.)

this last point. Particularly because these patterns of reduction are typically into what, from a human viewpoint at least, are increasingly marginal or unproductive lands, caution should obviously be exercised in interpreting anything about the physiological or ecological capabilities of species, for example for conservation purposes, on the basis of present patterns of distribution alone.

This said, there are a number of reasons to believe that climate plays a very significant role in determining the distributions of species. In isolation, none of these is overly convincing, but in sum they yield a reasonable case. First, climate matching approaches (sometimes extended to include other environmental variables) have also been applied to predict the likely occurrences of species when occurrence data are only available from parts of their distributions, as well as the likely distributional responses of species to future climate change (e.g. Walker 1990; Carpenter *et al.* 1993; Carrascal *et al.* 1993; Rogers and Williams 1994; Jeffree and Jeffree 1996; Romero and Real 1996). Such approaches have met with some success. Thus, for example, Robinson *et al.* (1997a,b) predicted the probable distribution of tsetse fly in unsurveyed areas by determining the environmental characteristics of areas of tsetse presence and absence in surveyed areas. With respect to single environmental variables, the best predictions for *Glossina morsitans centralis* were made using the average normalized difference vegetation index (75% correct predictions) and the average of the maximum temperature (70% correct), for *G. m. morsitans* and *G. pallidipes* the best predictions were given by the maximum of the minimum temperature (84% and 86% correct, respectively) (Robinson *et al.* 1997a). The predictive power of models improved from the simple to the more complex, and the use of a principal component analysis of multivariate climate and remotely sensed vegetation data followed by a maximum likelihood classification resulted in predictions for *G. m. centralis* and *G. m. morsitans* of 92.8% and 85.1%, respectively (Robinson *et al.* 1997b).

Typically, more than one, and often several, environmental variables make statistically significant contributions to the explanation of the pattern of occurrence of a species. Much attention has been paid to improving the statistical models used in such studies. Of course, hundreds of possible combinations of climatic parameters exist, and arguably it would be surprising if some combination could not be found which described the occurrence of a species reasonably well, regardless of any causal relationship.

Second, there is frequently geographical coincidence between the states of particular climatic factors (e.g. isotherms), and the distributional limits of species (e.g. Root 1988b). Root (1988a) investigated the association between environmental factors and distributional boundaries for 148 species of wintering land birds across North America. The comparisons revealed frequent correspondence between range limits and environmental factors. For example, average minimum January temperature, mean length of frost-free period, and potential vegetation were frequently associated with northern range limits. Less than 1% of all the associations observed were expected to occur by chance. It has been proposed that a physiological ceiling on metabolic rate, preventing the demands of thermoregulation from being met as temperature declines, constrains the northern distributional limits of these wintering birds (Root 1988a,b,c; 1989). However, subsequent analyses have challenged the contention (Repasky 1991), perhaps reinforcing the notion that limits in this case are likely to be determined by the interactions of temperature and biotic factors rather than simply by a physiological constraint.

Third, peripheral populations are often more susceptible to environmental change than core populations. Thus, Mehlman (1997) finds that for three species of passerine birds in North America proportional abundance changes following the harsh winters of the 1970s were greatest at sites closer to the range edge and, for two species, at sites with a more severe winter.

Fourth, distributional boundaries often seem to coincide with the onset of climatic conditions which are unsuitable for reproduction or the successful completion of life cycles. For example, Bryant *et al.* (1997) studied four nymphalid butterfly species that share the same primary host plant, the common stinging nettle *Urtica dioica*, but have different margins to their geographical ranges in the UK and mainland Europe. They found that the distributional margins appeared to follow summer isotherms, and that for three species, small tortoiseshell *Aglais urticae*, peacock *Inachis io* and comma *Polygonia calbum*, differences in relative distribution could broadly be explained in terms of degree-day requirements. The migrant red admiral *Vanessa atalanta* did not fit the predicted pattern, perhaps because it may be more limited by its ability to overwinter. The most northerly species, *A. urticae*, had the lowest degree-day requirement. In a similar vein, Strathdee and Bale (1995) found that the distributional limit of the high Arctic aphid *Acyrthosiphon svalbardicum* in an area of Spitsbergen was determined by summer thermal conditions. Aphids survive only at sites with little or no snow cover that have maximal summer thermal budgets, but are exposed to the lowest winter temperatures. They argue that in the high Arctic the distribution of many invertebrates is likely to be defined by the severe thermal conditions alone.

Fifth, outlying populations of some species appear to lie in climatic refugia. For example, Ryrholm (1988, 1989) documents the occurrence of the moth *Idaea dilutaria*, the beetle *Danacea pallipes* and the spider *Theridion conigerum* on the Kullaberg peninsula in southern Sweden, around 1000 km north of the northern limits of their continuous distributions. The area of the peninsula occupied by these species is distinctly warmer than surrounding and apparently similar habitats. This climatic outpost is a result of steep vegetated slopes, the relatively large number of sun hours in spring, proximity to the sea and the character of the rock outcrop.

The existence in climatic refugia of populations which are peripheral to the main body of a species distribution is associated with a more general observation that at the edge of their ranges many species tend to be restricted to specific habitats and microhabitats. For example, Hodkinson (1997) found that there was a progressive reduction of host plant species and tissues exploited by the willow psyllid *Cacopsylla groenlandica* towards high latitudes in Greenland, such that towards the northern limit of its range it became highly specialized, feeding only on female catkins of one host despite the presence of alternative hosts. In some cases, this greater specificity is likely to be associated with the occupation of microclimates which are less severe than conditions at large, although it could also reflect greater specific nutritional requirements when a population is at its margins.

The argument for the importance of climate in the direct limitation of the ranges of species has sometimes been put very forcefully. Thus, Brett (1956, p. 80) argued that 'There can be no doubt that the lethal temperature exerts limiting effects on the geographical distribution and freedom for successful existence . . . '. However, in interpreting the above observations, with regard to the precise roles played by climate and physiology in limiting geographical distributions, some circumspection is necessary. There are several reasons for this.

1 Most of the evidence is based solely on correlations between the occurrence of a species and the state of one or more climate variables, and cause and effect cannot be inferred from correlations.

2 Correlations between environmental factors and the occurrence of a species fail to demonstrate that the limits to its distribution coincide with those of its physiological tolerances and capacities. As already seen, individuals often seem to have tolerances greater than they require *in situ*: the realized physiological tolerance (what individuals *do* tolerate) may only be a subset of the fundamental physiological tolerance (what they *can* tolerate) (Sect. 3.4). One way round this problem is to use measurements of fundamental physiological tolerances from laboratory-based experiments to predict the likely geographical range for a given species. For example, Young (1941), in an attempt to correlate the distribution of the mussel *Mytilus californianus* with the salinity of the environment, used the salinity tolerances of adults and larva that he measured in the laboratory as a basis for predicting where this mussel should be. He found that there was a reasonably good fit between predicted and observed occurrences, but there were still some low salinity waters from which it was absent, even though he thought the mussels were 'suited to it.' Shine (1987) showed how data on thermal tolerances of embryos could be used to predict the distributional boundaries of oviparous and viviparous taxa of large venomous snakes of the genus *Pseudechis*. Howe and Lindgren (1957) (see also Howe 1958) attempted to predict how a species might spread, based on the distribution of potentially favourable conditions as determined in the laboratory. Obviously, a serious obstacle to such approaches is the difficulty of knowing a priori which, if any, physiological traits are likely to be of most significance in determining the geographical distribution of a species.

3 Correlations between environmental factors and the occurrence of a species could readily result if the environmental factors reflected the occurrence of a resource on which the species was dependent (e.g. a food plant, or prey species), or the occurrence of conditions under which a competitor, predator or parasite occurred in numbers low enough so as not to displace it (Cook 1931). Davis *et al.* (1998) illustrate the potentially marked effects of between-species interactions on the observed climatic envelopes of species, using an elegant set of microcosm experiments, a model assemblage of three fruitfly species, *Drosophila melanogaster, D. simulans* and *D. subobscura*, and a parasitoid wasp *Leptopilina boulardi*. Temperature clines were generated, by linking eight cages in series and housing each pair of cages in separate incubators at different temperatures. Competitive interactions between the fruitfly species altered their distributions and abundances from those found in single-species experiments. Adding the parasitoid further modified these distributions and abundances.

Field studies also emphasize the role of between-species interactions. For example, the high shore gastropod *Melaraphe* (= *Littorina*) *neritoides* has its centre of distribution in the northern Mediterranean but also maintains abundant and viable populations throughout the British Isles and has its northern range limit in Norway (Hughes and Roberts 1980). There are no effective competitors or predators of these northerly populations and it has been suggested that the low temperatures which they experience are close to the lower limit for successful reproduction in this species. Indeed, the case for abiotic factors, here temperature, limiting distribution looks convincing. Contrast this with the case of another member of the same genus. On the west coast of North America, the southern limit of the high intertidal *Littorina sitkana* is normally about 43°N, with individuals transplanted any further south succumbing to desiccation stress (Behens-Yamada 1977). However, Behens-Yamada (1977) points out that for transplanted populations there is a tremendous increase in predation pressure from crabs which are more abundant in the high intertidal zone in the south.

A role for competition in determining the range limits of a species is often inferred when its absence from certain regions cannot be attributed to its physiology. Thus, competition with a congeneric species was suggested as restricting the distribution of the southern African ice rat *Otomys sloggetti*, when its absence below altitudes of 2000 m was found not to be determined by its thermal physiology (Richter *et al.* 1997). *O. irroratus* and *O. sloggetti* have allopatric distributions, and Richter *et al.* (1997) argue that competition excludes the latter from exploiting lower altitudes, while thermoregulatory considerations preclude the former from exploiting higher ones. This combined effect of temperature and competition in determining species distribution is, they propose, a widespread one. The idea certainly has a distinguished history, with MacArthur (1972) proposing that geographical range limits at lower latitudes are more difficult to explain than high latitude limits, because the latter are more likely to be imposed by climate.

4 Even the demonstration that individuals cannot survive or cannot complete life cycles under climatic conditions beyond the edge of the range of a species remains in one sense only indirect evidence that those conditions are limiting. If other factors limit the occurrence of a species, then the potential to survive, develop and successfully reproduce under conditions beyond those limits may be lost, driving a tendency toward coincidence between conditions which are experienced and conditions which can be tolerated. Because physiologies must at least be sufficient to tolerate conditions within the range, this may result in systematic interspecific patterns of geographical variation in physiologies and the positions of geographical limits being observed (Chapter 5). The extent to which ecology determines physiology and vice versa is a much broader issue, to which we will return (Chapter 6).

5 In the context of predicting geographical range changes in the face of climatic change, which has become a widespread research activity, climate matching assumes (i) that any influence of other factors on observed relationships between climate and occurrence, such as resources, competitors and predators and parasites, will remain constant—additional results from the microcosm experiments of Davis *et al.* (1998) mentioned earlier show that, at least in the laboratory, this need not be so (see also Lawton 1995); (ii) that climate change will be relatively simple, inasmuch as its influence on species distributions can be summarized in terms of the projected changes in one or a few variables—in fact, several components of climate are expected to change, and the responses of animals may be sensitive to interactions between these components; there need be no exact analogues of present climates in the future (Lawton 1995); (iii) that there is no physiological capacity to withstand environmental conditions which are not components of those prevailing (see Sect. 4.5.2)—as we have reiterated several times, this is not necessarily so; and (iv) that range shifts, expansions or contractions (or introductions for that matter) are not accompanied by physiological changes, other than local non-genetic physiological acclimatization—it would not seem unlikely that, faced with a novel set of environmental conditions, local genetic adaptation could occur.

   Climate is doubtless of fundamental significance in determining the limits to species distributions. Demonstrating when, how and why is, however, seldom straightforward.

4.5.2 Why don't species have larger geographical ranges?

Whether geographical distributions are limited by climatic conditions or not, the question arises as to what prevents evolution from occurring such that a species can expand its range. This is particularly pertinent where no obvious barriers to expansion exist (such as physical barriers or the absence of a critical resource). Several hypotheses have been proposed to explain this conundrum (Mayr 1963; Hoffmann and Blows 1994; Hoffmann and Parsons 1997; Kirkpatrick and Barton 1997). While we do not propose to evaluate them all here, the first three concern low heritabilities of traits determining range boundaries in marginal populations: (i) low overall levels of genetic variation occur in marginal populations because of small population size; (ii) traits show low heritability as a consequence of directional selection in marginal environments; and (iii) traits show low heritability because of environmental variability in marginal environments. A further three hypotheses concern genetic interactions between traits: (iv) changes in several independent characters are required for range expansion, and so favoured genotypes occur too rarely; (v) genetic trade-offs between fitness in favourable and stressful environments prevent the increase of genotypes adapted to stressful conditions, because conditions limiting species margins occur infrequently; and (vi) genetic trade-offs among fitness traits in marginal conditions prevent traits from evolving. Two other hypotheses are that: (vii) the accumulation of mutations deleterious under stressful conditions prevents adaptation; and (viii) genotypes favoured in marginal populations are swamped by gene flow from central populations.

There is no generally accepted conclusion as to which of these hypotheses is most appropriate, and in what circumstances, and an examination of the evidence would extend well beyond the subject matter of this book. However, if the last hypothesis, which is presently gaining some currency, is correct then most of the between-population differences seen in physiological processes should be the result of local physiological acclimatization, as most of the genes in peripheral populations are typically adapted to the conditions at the centre of the range. This concords with a long-held view that for widely distributed species, physiological traits are approximately the same throughout their geographical range (e.g. Ushakov 1964), with any observed differences being due to local physiological acclimatization.

4.5.3 Do species escape climatic constraints on their ranges?

The vast majority of observed changes in the distributions of animals, which are not obviously mediated by the effects of introduced species or habitat change/destruction, are interpreted, probably correctly, in terms of the direct or indirect influence of changes in prevailing climatic conditions. The covariance frequently observed between changes in the distribution of a species and changing climatic conditions, usually temperature, may well often be, at least proximally, causal (e.g. Murawski 1993; Burton 1995; Parmesan 1996; Whittaker and Tribe 1996). Murawski (1993) regressed the mean and maximum latitude of occurrence of 36 fish and squid species in the north-west Atlantic against average surface- and bottom-water temperatures. Between-year variations in water temperature were significant in explaining changes in mean latitude of occurrence for 12 species in spring and autumn. Responses in maximum latitude to interannual differences in temperatures occurred for five species in both spring and autumn.

Burton (1995) argues that changes in climate since the beginning of the 20th century have probably, at least in part, resulted in at least 309 of the 435 European breeding bird species undergoing alterations in their geographical distributions. Of these, 224 species have moved northward, north-westward and/or westward, 32 species have retreated south or south-eastward. Between 1900 and 1950, 50 species retreated northwards and since then 55 species have advanced southwards again.

The possibility remains, however, that, all else being equal, a species may escape the climatic constraints on its distribution through a change in the genetics controlling its physiology. Observing such cases is not easy. The problem is elegantly expressed by Lewontin and Birch (1966, p. 315):

'How are we to distinguish such preadapted ecological shifts from cases of real change in the adaptive norm of a species? We might suspect the latter was happening if long established boundaries of its distribution began to move despite lack of change in relevant components of the environment in the invaded area. Of course, we may never be certain that some subtle but relevant change has not occurred. However, this is a limitation that all studies of ecology and genetics of natural populations suffer from. One can never be certain he has exhausted the relevant aspects of the environment. One can only be critical and hope.'

These authors go on to document a possible case of range expansion through the generation of a novel physiology. The fruitfly *Dacus tryoni* has, since at least 1860, extended its geographical range southward in eastern Australia. There has also been a concomitant increase in altitudinal range. It was suggested by Lewontin and Birch (1966) that this broadening of geographical range was limited by climatic factors rather than by biotic ones (e.g. availability of ripe fruit required for egg laying). Laboratory investigations revealed that range extension could be correlated with wider heat and cold tolerances, with the genetic variation required for selection apparently being provided by introgression ('coming in') from a second species of *Dacus*, *D. neohumeralis*. Unfortunately, hybridization of both *Dacus* species in the laboratory resulted in the production of hybrid populations better adapted to only some of the experimental temperatures used. Despite some of its obvious weaknesses, this study is one of the few in which genetically mediated physiological change appears to result in changes in geographical range. Both in the case of clinal variation in physiology (Sect. 4.4) and the production of genetic/physiological novelty which allows range extension (above), it is perhaps reasonable to suggest that the selection of genetic variants may not only alter the range of the species but may also be involved in the process of speciation itself. If this is indeed the case, it may be that environmental temperature plays an important role in this process.

4.6 The link to fitness

Implicit in much of the preceding discussion, and in the literature at large, is the assumption that between-population variation in physiological traits serves to enhance the fitness of individuals in different populations. The evidence that this is actually so remains sparse. Most attention has been directed at the link between physiological variation and fitness at the between-individual level (Sect. 3.6). In the most general sense, reviewing the available evidence, Hoffmann and Blows (1993) conclude that populations marginal to the range of a species may often display low fitness except under marginal conditions, at least where there is evidence of ecotypic differentiation.

A few studies bear directly on between-population variation in a physiological trait and fitness. Thus, Ayres and Scriber (1994) find that, on a good host plant, at Alaskan temperatures, individuals of the Canadian tiger swallowtail *Papilio canadensis* from Alaska had an estimated fitness three times greater than individuals from Michigan; the Michigan population was predicted to become extinct in 31 of 48 years at Alaskan temperatures. Four differences between Alaskan and Michigan populations are interpretable as adaptations to short, cool subarctic summers: (i) increased egg mass; (ii) reduced adult size; (iii) enhanced moulting abilities at low temperatures; and (iv) enhanced growth rates at low temperatures.

Likewise, there are pronounced population differences in neuromodulatory mechanisms associated with predator recognition and escape responses in a mainland and an island population of deer mice *Peromyscus maniculatus artemisiae* (Kavaliers 1990). In a weasel-sympatric mainland population, exposure to 'weasel scent' elicits an immediate escape response; this escape response is not so pronounced in individuals of an island (predator-free) population.

Adult flies of *Drosophila mojavensis* are exposed to fermentation substances (a presumed energy source for these flies) present in 'rots', of cacti endemic to the deserts of Mexico and Arizona (USA). Starved individuals exposed to ethanol vapour, or in the presence of fermenting cacti in the field, live longer, are more fecund and have greater metabolic rates, than those exposed to water vapour (Etges and Klassen 1989). The authors suggest that the ability to survive and reproduce in low molecular-weight, volatile-rich, carbohydrate-poor cactus environments has allowed *D. mojavensis* to colonize extensive desert regions now within the range of the species.

In general, it remains possible that (i) physiological diversity has fitness implications; and (ii) the existence of physiological variation, whatever its origin, serves to reduce fitness variation between populations. However, much work remains to be done in this area. Almost certainly, this investigation of the fitness implications of between-population variation will necessitate the use of more direct measures of fitness.

4.7 Concluding discussion and summary

'There was an exuberant fierceness in the littoral here . . . starfish and sea urchins were more strongly attached than in other places . . . Perhaps the force of the great surf which beats on this shore has much to do with the tenacity of the animals here.'
[Steinbeck, 1990, p. 120]

Physiological differences between populations of the same species are readily observed. They result, to varying degrees, from acclimation (reversible and non-reversible), genetic differentiation and differences in demography. They can be systematic across space, constituting broad geographical patterns with respect to latitude, altitude and depth. However, the nature of the patterns and their underlying mechanisms are still equivocal in many cases. Some work has been carried out on the relationship between range size and physiological variation, although much of what is known is still to a large extent conjecture. Sometimes, as a result of inhabiting locations characterized by similar environmental conditions, an absence of the capacity for local acclimatization, or because of significant gene flow, different populations have broadly the same characteristic physiology.

As yet, and somewhat depressingly, there are too few case studies, particularly carried out in a comparable fashion, to enable generalizations to be made about when and where between-population differences in physiological capacity are most likely to be encountered. The fact that such variation can occur and that, at least in some cases, it is greatly in excess of within-population variation, means that variation at this level of organization cannot be ignored. It is at this level, perhaps, that environmental and genetic effects are most readily differentiated, although care needs to be taken in the construction of experiments.

Chapter 5: Species Contrasts

'If physiology is to be a true science, it must recognize the diversity of animal life before it claims generalizations based on a limited selection of convenient animal types.'
[Florey, 1987]

5.1 Introduction

Ask most people for an example of physiological variation and it is unlikely that they will suggest one which concerns variation within individuals, between individuals within populations, or between populations. Rather, in the vast majority of cases they will suggest an example of variation at the next hierarchical level, and the final one which we will consider in this book, namely that between species. This should not be surprising, if for no other reason than that the breadth of physiological variation between species is, at least in the extreme, so much greater than that within a species. No species is physiologically capable of surviving from the tops of high mountains to the depths of the oceans and from the highest latitudes to the lowest, and the variations species exhibit in their distribution are all too readily linked to variations in physiology.

Although physiological variation at all levels has important ecological implications (Chapters 2–4), it is at the level of between-species comparisons that this has most overtly been recognized. This is evident when perusing almost any textbook of comparative animal physiology or ecological physiology. It is also particularly true with regard to many of the broad geographical-(large-) scale issues with which ecologists have been much concerned. That is, issues such as why most species occur in the tropics (Pianka 1966; Rohde 1992; Rosenzweig 1992, 1995; Gaston and Williams 1996), why species richness declines towards high elevations (Rahbek 1995; Gaston and Williams 1996), why most species occur at low local abundances and are narrowly distributed (Preston 1948; May 1975; Hanski 1982; Tokeshi 1993; Gaston 1994; Brown 1995; Gaston 1996a,b, 1998; Gaston *et al.* 1997), and why most are small-bodied (May 1978; Maurer *et al.* 1992; Brown *et al.* 1993; Blackburn and Gaston 1994; Kozlowski and Weiner 1997). These are all fundamental questions about the distribution of life on Earth, which must be addressed if ecology is genuinely to be the study of the abundance and distribution of organisms (Krebs 1972). More recently, and for more pragmatic reasons, other large-scale issues have come to the fore, the answers to which again necessitate some understanding of between-species differences in physiological tolerances or functions for numbers of closely related species (taxonomic assemblages). These include questions such as how best to maintain biodiversity in the face of competing land-use, how species will respond to global environmental change, and how best to predict the course and outcome of species introductions (for reviews see Peters and Lovejoy 1992; Kareiva *et al.* 1993; Gaston 1996c; Williamson 1996).

To tackle all of the issues raised above, even in brief, would require the writing of a substantial tome. Consequently, in what follows we have had to be more selective in the ground covered compared with previous chapters. In this chapter, we consider the nature of physiological variation between species. However, to avoid retracing many of the well-worn paths explored in textbooks of ecological physiology and comparative animal physiology, we will focus primarily on those patterns of physiological variation which seem of most relevance to geographical-scale ecological issues. This means we will be principally concerned with patterns of systematic variation with respect to space. In doing so we parallel similar sections on latitude, altitude and depth as presented in the context of between-population comparisons in the previous chapter. We begin, however, by considering the sources of physiological variation between species.

5.2 Sources of between-species variation

Between-species variation in physiological traits can derive from a number of sources. Significantly, in broad terms, in many cases these closely parallel those sources which generate between-individual variation in physiological traits (Sect. 3.5).

5.2.1 Measurement and summary statistics

The least interesting source of between-species physiological variation is doubtless differences in the methods which have been used to measure a particular physiological trait. Broadly, four kinds of between-species studies can be recognized. First, there are those actually studying single species, or perhaps a couple from the same location, for which the physiological measurements made are compared with literature values for species from other locations. The key problem here lies with comparability between studies. Often they are not directly comparable, due to the use of different experimental protocols or differences in the physiological condition or behaviour of the individuals studied. Furthermore, even if care is taken to replicate experimental protocols used in older studies, often the techniques or apparatus employed today have been subject to considerable development in the intervening time.

Second, and perhaps more often, investigators choose to study two species each occurring at a different location, often, respectively, at high and low latitudes or altitudes or depths (note the predominance of two species studies throughout this chapter). Here, the problem is that the investigator is assuming that any differences detected can be accounted for by the difference in latitude, altitude or depth, and not another factor, such as irreversible acclimatization, phylogeny or some other ecological or geographical difference (cf. Garland and Adolph's (1994) critique of inferring adaptation from two-species studies). Comparison of these data with others from the literature is subject to the same criticisms as can be levelled at single-species studies.

Third, some investigators attempt to draw on derived data from as many sources, and often for as many species within a taxon, as possible. This approach is one in which the investigator has no control over the original experimental designs and mode of data collection. The only thing they can do is to be selective in which studies to use and which to omit. This too is subject to the same criticisms levelled at the previous two approaches. That said, and notwithstanding all that has gone before, such an approach may, for whatever reasons, still have discriminatory power. Thus, drawing on about 25 different studies, Degan *et al.* (1998) collated data for a total of 63 species of rodents, and found significant and probably real relationships between metabolism and body size. This, despite original data collection having been carried out under different experimental protocols and conditions (e.g. air temperature range 10–25°C).

Finally, a far smaller number of investigators have directly studied a significant range of different species, particularly across a reasonable spread of locations or environmental conditions. This approach is preferred as it addresses many of the problems of comparability. However, there are still some potentially significant problems to address. For instance, comparisons of physiological traits for a number of species tend to be based on mean values across typically rather few individuals. This assumes that these values are representative of the physiologies of the different species. The preceding three chapters have emphasized just how variable within-species differences in physiological traits (at every level) can be, suggesting that the assumption might frequently be violated. Where within-species variation is explicitly considered, if this is determined in different ways for different species, artefactual between-species differences in physiological traits may be generated. This is a particular problem when considering the relationship between the breadth of physiological tolerance of a species and, say, its abundance or geographical range, where differences in sample sizes (large for common and widespread species, and small for rare and restricted ones) may undermine the interpretation of comparisons.

5.2.2 Phylogenetic relatedness

Most of the literature concerned with patterns of between-species variation in physiological traits is based on sets of trait values for each species under consideration. It seems most natural, therefore, to use the species as the unit of analysis. This assumes that species can be treated as independent data points.

Because of their phylogenetic relatedness this is not necessarily so, and the degrees of freedom available for testing statistical significance may be inflated (Harvey and Pagel 1991; Harvey 1996). If sufficient, this inflation may falsely imply patterns of variation between species which in reality do not exist. Some of the clearest examples of such problems derive from analyses of morphological, life history and ecological traits. For example, Nee *et al.* (1991) showed that the negative relationship between abundance and body mass in British bird species resulted from a difference between passerines, which tend to be small-bodied and common, and non-passerines, which tend to be large-bodied and rare. Within each group there was no evidence for any association between abundance and body mass. Therefore, this relationship results from a single evolutionary difference between passerines and non-passerines, rather than any general tendency for abundance and body mass to be negatively related.

A number of statistical methods have been developed for testing hypotheses about continuous-valued characters, of the form of the physiological traits we have addressed in this book, while controlling for the effects of phylogenetic non-independence (e.g. Harvey and Pagel 1991; Gittleman and Luh 1992; Miles and Dunham 1993). Independent contrasts is one of the most widely used and best understood of these approaches. First proposed and used by Felsenstein (1985), the method has been explored and refined by a number of subsequent workers (Grafen 1989; Harvey and Pagel 1991; Garland 1992; Purvis and Garland 1993; Losos 1994; Purvis and Rambaut 1995). The use of independent contrasts when examining physiological data has, over the past few years, become a more frequent feature of studies in ecological physiology (e.g. Chown *et al.* 1997; Degan *et al.* 1998). This said, it is in the study of the evolution of physiological traits that such analyses have had perhaps the greatest impact. While ecological physiologists for many years have claimed to be studying evolution, their efforts had gone largely unrecognized by evolutionary biologists (Prosser 1960; Burggren 1989; Burggren and Bemis 1990). However, the application of phylogenetic analyses allowing testable predictions on the evolution of such features as endothermy, aerobic capacity, metabolic rate, specialized nerve cell function and mammalian blood characteristics, has to a large extent invigorated what has come to be known as evolutionary physiology (Elgar and Harvey 1987; Derrickson 1989; Garland *et al.* 1991, 1997; Diamond 1992, 1993; Garland 1992; McNab 1992; Walton 1993; Clark and Wang 1994; Garland and Adolph 1994; Garland and Carter 1994; Grueber and Bradley 1994; Ruben 1995; Diaz *et al.* 1996; Dutenhoffer and Swanson 1996; Ward and Seely 1996; Wright *et al.* 1996; Alves-Gomez and Hopkins 1997; Bennett 1997; Mangum and Hochachka 1998; Weibel *et al.* 1998); of course, this is not to say that the field of evolutionary physiology is only concerned with between-species comparisons.

Nevertheless, there are relatively few studies which investigate phylogenetic non-independence in the context of the implications of physiological diversity for ecology. Consequently, it is still unclear how significant it actually is. On the one hand, it seems likely that some physiological traits exhibit rather little phylogenetic constraint, such that species can justifiably be treated as independent data points for analysis. For example, a correlation between basal and maximum (cold-induced) metabolic rate for 10 passerine species was found to exist, whether or not phylogeny was accounted for (Dutenhoffer and Swanson 1996), as did a negative relationship between warm-up rates and the minimum temperature for activity in bees (Stone and Willmer 1989). On the other hand, there is seldom sufficient a priori knowledge to be able to reject a possible role for phylogeny and thereby avoid explicitly accounting for it in analyses. Moreover, with regard to between-species variation in physiological traits, interest is usually focused on how these traits covary with other variables. In the case of variables of ecological interest, these often themselves exhibit moderate to strong phylogenetic non-independence, necessitating that relatedness be taken into account when conducting analyses. Adult body size, for example, exhibits marked phylogenetic conservatism.

A few studies find that controlling for phylogenetic relatedness serves to reveal patterns which were entirely obscured when species were treated as independent data points. Thus, Promislow (1991) studied taxonomic and allometric differences in blood parameters for 206 species of mammals. Contrary to previous studies, this work found that body mass was a significant predictor of key blood parameters. In the study on the temperature relations of bees, referred to above (Stone and Willmer 1989), a correlation between warm-up rate and body size was only demonstrated when controlling for phylogenetic relatedness (cf. Hayssen and Lacy 1985; Elgar and Harvey 1987; Derrickson 1989).

The general lack of studies of physiological diversity in an ecological context which control for phylogenetic non-independence means that considerations of this topic have to be founded on other studies. Interpretations and conclusions drawn from these in later parts of this chapter must therefore be regarded as provisional; a reanalysis of existing work to redress the situation is a substantial task.

Although arguably at its most severe at the level of between-species comparisons, the complications resulting from the non-independence of data points for analysis as a result of relatedness extend to lower levels in the hierarchy of physiological variation. At the levels of between-individual variation, both within and between populations, individuals will be genetically more closely related to some conspecifics than to others (see Sects 3.5.3, 4.2.4). It would therefore be desirable to control for such effects in any analyses which are conducted, although we are not aware of studies which have actually sought to do this, and it will usually be difficult to achieve.

5.2.3 Species circumstances

Differences in the circumstances in which individual animals live was identified as a possible source of between-individual physiological variation for particular species (Sect. 3.5.2). Differences in the circumstances inhabited by different species may similarly provide a source of variation in physiology between individuals of those species. Indeed, in principle, this remains a possible explanation for between-species variation whenever physiological traits are measured for different species under different conditions. The general occurrence of acclimation (or acclimatization in cases where animals are taken straight from the field), particularly when it is irreversible (Sects 2.4.1, 4.2.3), means that physiological differences may be attributable to this source alone. Depending on the rationale for a study, whether acclimation is the source of physiological variation may or may not be relevant. Indeed, for many ecological studies the inclusion of acclimatory effects is arguably of more interest than their exclusion. It is not uncommon that the conditions in which animals are held in the laboratory dictates the outcome of any physiological patterns observed. For example, when comparing the thermal tolerances of two species of snake, while the mean upper thermal tolerance (CT_{max}) for *Natrix rhombifera* was significantly greater than that for *Thamnophis proximus* when both are kept at 15°C, keeping them at 30°C reverses this pattern (Jacobson and Whitfield 1970) (cf. the range of 'acclimation outcomes' for whole animal and tissue oxygen uptake of a number of fiddler crab *Uca* species belonging to a number of different populations—Vernberg and Vernberg (1966a,b)). As yet, too few studies comparing species have incorporated acclimation treatments into experiments or tackled the issue of non-reversible acclimation. Consequently, it is still difficult to be categoric with regard to the proportion of the total variation observed at the between-species level that can be accounted for by non-genetic factors (but see Sect. 6.2).

5.2.4 Genetic differentiation

The physiological variation between species which is of greatest interest, although not necessarily the most prevalent (Sect. 5.2.3), is that which has been generated by genetic differentiation. In addressing this issue, we begin by considering how physiological variation and speciation may interact.

Physiological differentiation, hybridism and speciation

On a number of occasions, Prosser (1957b, 1964b, 1986) expressed the view that as physiological variation provides a basis for natural selection, knowledge of such variation is critical to an understanding of speciation. He suggested that a full description of the adaptive physiology of a species, with respect to niche and geographical range, would provide a functional definition of a 'physiological species', in terms of isolating mechanisms (geographical and behavioural: Prosser 1960, 1964b, 1986). Such a scheme potentially allows recognition of speciation even in the absence of morphological difference. Currently, there are at least seven distinct species concepts (Bisby 1995), and some are close to the notion of a species as set out by Prosser. If accepted, the physiological species concept would necessitate the recognition of many of the between-population differences presented in the previous chapter actually as species differences. Generally, it is found that, for most purposes, working with the morphological species concept is sufficient and, indeed, in many cases is the only concept which is practicable. Given this constraint, the notion of a physiological species may not be a helpful way forward, although clearly in some cases where substantial physiological variation is detected, the question of whether the two populations actually constitute different species should be considered closely.

Speciation and physiological diversity can be correlated with ecological heterogeneity. The distribution patterns of four chromosomal species (a species complex) of the subterranean mole rat *Spalax ehrenbergi* are associated with four climatic regions in Israel: cool and humid ($2n = 52$); cool and dry ($2n = 54$); warm and humid ($2n = 58$); and warm and dry ($2n = 60$). The dry regions are characterized by lower productivity than the humid ones and so there is less food available to the mole rats in those regions. Consequently, there is species divergence along the lines of aridity and habitat productivity. The complex shows xeric ($2n = 54, 60$) and mesic ($2n = 52, 58$) adaptations in diverse characteristics, molecular (at the level of proteins and DNA) and whole organismal (physiological, morphological and behavioural). Physiological diversity manifests itself in metabolism and respiration (Nevo and Shkolnik 1974; Arieli *et al.* 1984, 1986; Yahav *et al.* 1988a), thermoregulation (Haim *et al.* 1984, 1985) and water balance (Yahav *et al.* 1988b; Nevo *et al.* 1989).

Genetic and physiological differentiation have been studied in two closely related marine mussels and their hybrids (exactly how close is the relationship is still unclear, e.g. Varvio *et al.* 1988). *Mytilus edulis* occurs in temperate cold waters of North America and Europe, and *M. galloprovincialis* occurs in the Mediterranean and has its northern limit in France and the UK (Hilbish *et al.* 1994). In particular, attention was paid to divergence in the effects of temperature on physiological energetics, because of the seemingly different capacities of the species to occupy separate biogeographic regions. *Mytilus* populations in the UK are a mosaic of possible combinations of the species (Skibinski *et al.* 1983), and the particular population studied by Hilbish *et al.* (1994) consists both of F_1 hybrids and a number of introgressed genotypes. The two species are physiologically differentiated and, under warm-water conditions (23°C), these differences result in *M. galloprovincialis* having a greater net energy balance than *M. edulis*; the former possessed a greater rate of feeding and metabolism than the latter, which could be correlated with an elevated growth rate *in situ*. This study, taken together with the biogeographical data and a study which shows that *M. edulis* may have a growth advantage at low temperatures (Seed 1971), suggests that *M. galloprovincialis* is physiologically more capable than *M. edulis* at exploiting environments characterized by high summer temperatures. Furthermore, two of four isozyme markers that show high degrees of differentiation between the species (esterase and octopine dehydrogenase), and may be linked and cosegregate with physiological differentiation, explain most of the physiological differences between them. Hilbish *et al.* (1994) conclude that these results suggest that the physiological differentiation they observe is controlled by one, or at least a very few, genes.

The study of hybrids has resulted in advances in our understanding of what is, or may be, happening during speciation. A number of such studies have investigated at molecular and whole organism levels the possibility of physiological heterosis, where most of the values for a particular characteristic lie outside of the range of the means for both of the parent populations; the effect can be positive (hybrid vigour) or negative. Manwell *et al.* (1963) investigated the possibility that hybrids of fish or bird species might produce a novel haemoglobin molecule as a result of alterations in the constituent polypeptide chains. For most of the crosses they performed there were no new proteins, and the oxygen-binding properties of the hybrid haemoglobin were identical to those of a mixture of the two species used in the cross. However, on crossing Warmouth sunfish *Chaenobryttus gulosus* with green sunfish *Lepomis cyanellus*, the resultant hybrid possesses a haemoglobin that is quite distinct, structurally and functionally, from either parent species or a mixture of their haemoglobins (Fig. 5.1). In terms of function, the hybrid haemoglobin is better suited to providing for a higher rate of metabolism (hybrids are more energetic and aggressive than either parent species). Unfortunately, these investigators were unable to show conclusively that the hybrid haemoglobins were the result of polypeptide chain recombination.

Often very small differences in physiology accompanying speciation can have profound and far reaching implications by dramatically extending the range of environments available to the new species. Amphibian, bird and mammal species that live at high altitudes possess haemoglobins with a greater affinity for oxygen than their low altitude counterparts. The increase in oxygen affinity is, it has been suggested by Perutz (1983), the result of one (or a few) amino acid substitution in a key position, the majority of substitutions that characterize species differences being functionally neutral. For example, the Andean goose *Chloephaga melanoptera* lives at 5000–6000 m altitude in South America all year round, and the bar-headed goose *Anser indicus* lives and hatches its young at 4000–6000 m altitude in Tibet and then migrates across the Himalayas (they have even been observed flying over the summit of Mt Everest) to the plains of north-west India. Both possess haemoglobins with a higher affinity for oxygen compared with, say, the greylag goose *Anser anser* which inhabits the plains of India all year round (Hall *et al.* 1936; Petschow *et al.* 1977). The difference in haemoglobin structure between the Andean and bar-headed geese and the greylag is small (16 and 4 mutations, respectively) with, in the comparison of bar-head and greylag for example,

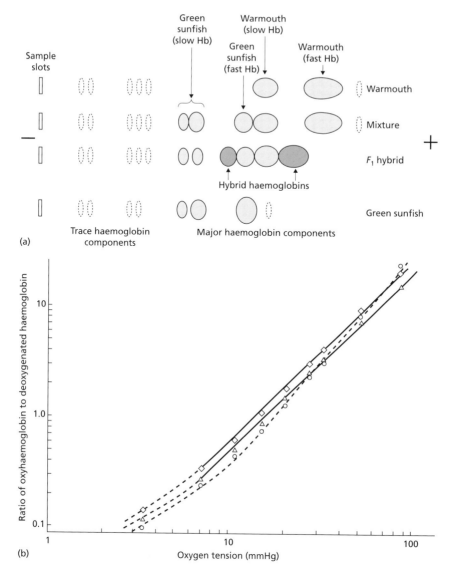

Fig. 5.1 Structure and function of haemoglobin from Warmouth sunfish *Chaenobryttus gulosus*, green sunfish *Lepomis cyanellus*, and their F_1 hybrid. (a) Diagrammatic representation of haemoglobin subunits separated using starch gel electrophoresis. (b) Oxygen-binding by haemoglobin from Warmouth sunfish (diamonds), green sunfish (triangles) and hybrid (circles). (After Manwell *et al.* 1963, with permission from Elsevier Science.)

only one mutation of any functional significance (i.e. the replacement of the amino acid proline with alanine at a key part of the globin molecule leaves a 2-carbon gap). A similar situation is seen in a comparison of Andean and greylag geese, and such gaps have been correlated with an increase in oxygen affinity. An increase in affinity is advantageous at high altitudes where lack of oxygen is a problem. In fact, Jessen *et al.* (1991) tested Perutz's idea by genetically engineering two human haemoglobins so that proline is substituted for alanine at a key location. This results in an increase in oxygen affinity, the magnitude of which was slightly greater than the affinity difference observed between bar-head and greylag geese haemoglobins. In conclusion, the increased affinity for oxygen of haemoglobins from high-altitude geese, which seems to be one of the key factors in the ability to fly at altitude (Faraci 1991; Weber 1995), is associated (in at least two species) with a single mutation (i.e. substitution of proline with the 'smaller' alanine). This relatively minor change in biochemistry and physiology (evolved separately in two geese species that inhabit different parts of the world!) results in the ability to exploit ecological niches that other geese species cannot tolerate.

The notion that physiological variation between species must always be greater the more distantly related they are, is an attractive one. It sits well with the view that the level of phylogenetic diversity tends to encompass not only the degree of relationship, but the degree of difference in many other characteristics too (Williams and Humphries 1996). However, it does not sit comfortably with the frequent occurrence of convergence, of which the pattern of haemoglobin–oxygen binding in the two geese species discussed above is a good example. Equally striking is the remarkable similarity of the function (though not the structure) of the respiratory pigment haemoglobin in terms of reversible oxygen binding in groups as disparate as vertebrates, echinoderms, molluscs, crustaceans, annelids, nematodes and even flatworms (Dickerson and Geis 1983; Hardison 1998; Terwilliger 1998), and the presence of a near identical anaerobic capacity involving the generation of L-lactate as an end product in crustaceans and vertebrates. While it may be objected that such convergence is likely to be more common at the molecular as opposed to the whole organism level (although how true is this?), there are whole organism examples of convergence as well. These include the ability of a number of aquatic species (vertebrate and invertebrate) to maintain oxygen uptake in the face of acutely declining oxygen tensions (Lutz and Storey 1997), and the ability of most freshwater species, regardless of ancestry, to maintain the osmotic concentration of their body fluids above that of the surrounding medium (Rankin and Davenport 1981; Gilles and Delpire 1997).

Somewhat surprisingly, there are also a considerable number of examples of strongly divergent evolution of physiological traits even among very closely related species (or even subspecies). They too render insecure such a general assumption of physiological diversity increasing with phylogenetic distance. For example, members of the amphipod genus *Gammarus* can be found in freshwater, estuarine, marine and hypersaline environments. The patterns of osmoregulation found in these closely related species encompass nearly all of the major patterns possible (Fig. 5.2; also Beadle and Cragg 1940; Werntz 1963; Lockwood and Inman 1973; Dorgelo 1977). The predominantly marine species (although they can also be found in brackish waters) *G. finmarchicus*

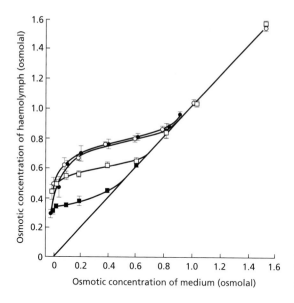

Fig. 5.2 Osmotic concentration of the haemolymph as a function of the osmotic concentration of the medium for four *Gammarus* species: *G. finmarchicus* (open circles), *G. oceanicus* (filled circles), *G. tigrinus* (open squares) and *G. fasciatus* (filled squares). (After Werntz 1963, with permission from the Editor of the *Biological Bulletin*.)

and *G. oceanicus* both maintain the concentration of their haemolymph at the highest levels and exhibit hyperosmotic regulation even at relatively high salinities. The freshwater *G. fasciatus* maintains the concentration of haemolymph at the lowest levels and is only able to hyperregulate over a much lower range of salinities than the marine forms. The predominantly brackish water *G. tigrinus* is intermediate to the marine and freshwater species. The subspecies of another *Gammarus* species, *G. duebeni,* can be found in both brackish and fresh waters (Hynes 1954) (cf. the example of a subspecies of marine crab which has colonized a freshwater lake (Forbes and Hill 1969)).

The possibility of such divergence even in closely related species is the implicit basis of some works on comparative animal physiology which examine ecological physiology with reference to different systems, for example, osmotic and ionic regulation (Potts and Parry 1964; Rankin and Davenport 1981; Gilles and Delpire 1997); nitrogen excretion (Kirschner 1967; Wright 1995); digestion (Stevens and Hume 1996); nervous function (Bullock and Horridge 1965; Nilsson and Holmgreen 1994); and metabolic arrest (Hochachka and Guppy 1987). It is suggested that the possibility should be entertained that, at least in some cases, convergent evolution in distantly related species and divergent evolution in closely related ones is much greater in physiological traits than might be found in morphological ones. In other words, close relatives, say two species of crab, are not necessarily more likely to share the same particular physiological attributes than are distant relatives, say a species of crab and a species of snail. If this is the case then accounting for phylogeny should matter less for physiological, as compared with mor-phological, traits.

Speciation is not inevitably accompanied by physiological differentiation, and sometimes the physiology of a particular ancestral species still appears to dominate. The desert pupfish *Cyprinodon nevadensis* and related species and subspecies, can be found in rivers, streams and thermal springs in one of the hottest places on Earth, Death Valley, southern California (USA). They are thought to have derived from an ancestral stock that invaded Lake Manly, which filled Death Valley in the early Pleistocene, with populations being isolated in permanent springs and streams at different intervals as the water level responded to climatic conditions. All of the populations of pupfish examined by Brown and Feldmeth (1971), from six thermally constant springs and three thermally fluctuating streams or marshes, are tolerant of a wide temperature range ($\Delta T = 40°C$) and show a pronounced acclimatory ability, even those populations inhabiting thermally stable environments. Similarly, high altitude anuran species from the geologically recent mountains of Central America are, in terms of their acclimatory ability, essentially the same as lowland tropical species. Brattstrom (1968) suggests that the high altitude forms are lowland forms that have been carried or forced into a variety of restrictive physiological plasticities. In contrast to what was said in the previous paragraph, these studies emphasize the importance of taking phylogeny and historical contingency into account when performing comparative studies. They also challenge the notion, often taken as axiomatic by biologists, that animals cannot do better than where they are at the present.

Physiological heterochrony

We have seen previously that functions or regulations which appear at a particular time in the life of one individual can appear at different times, sometimes not at all, or in different sequence, in other individuals of the same species (Sect. 2.4.4). Such differences in timing are also characteristic of comparisons between species, providing a source of between-species physiological variation; interspecific heterochrony is at the centre of current interest in the interface between development, evolution and ecology (Reilly *et al.* 1997). Therefore, the heterochrony of physiological traits in individuals of different species should also be of interest to ecological physiologists. One of the main considerations is the extent to which observed interspecific heterochrony is actually genetically fixed and how much it is open to, or even generated by, environmental influences (see Sect. 2.4.4).

Presented in Fig. 5.3 is the development of cardiac function in three crustacean species. Each possesses a quite different morphology and ecology from the other two. While the pattern of heart function is very similar in each of the species, the timings of the key events are not. In *Artemia franciscana* the onset of heart function is in the free-living posthatch state, whereas in the waterflea *Daphnia magna*, while the onset is posthatch, the hatchlings are still within the mother's brood pouch (marsupium). In common with most crab and lobster species, the heart beat of the amphipod *Gammarus duebeni* commences well

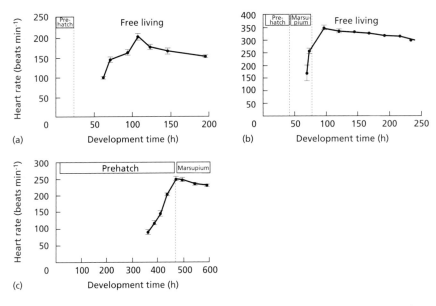

Fig. 5.3 Onset of cardiac activity and subsequent changes in heart rate with time in (a) the brine shrimp *Artemia franciscana*; (b) the waterflea *Daphnia magna*; and (c) the amphipod *Gammarus duebeni*. All values are means ± 1 SD. (*n*=10–14). (After Spicer and Morritt 1996, with permission from Elsevier Science.)

before hatching. Given the fact that heart development is so tightly constrained by morphological development (cf. Sect. 2.4.4), it is extremely likely that what are perceived as species differences in the timing of the ontogeny of cardiac function are genetically determined and cannot readily be explained on the basis of within-individual heterochrony.

Compare this with differences between three species of anadromous salmonids in the timing of (and 'trigger' for) the onset of the physiological regulations associated with the 'switching on' of salinity tolerance (McCormick 1994). In the brook trout *Salvelinus fontinalis*, the ontogeny of salinity tolerance is primarily triggered by exposure to salinity and occurs later in development than in the Atlantic salmon *Salmo salar*, where the primary trigger is photoperiod (Fig. 5.4a,b). In both cases, while there is an itinerary for physiological development, this is very much open to environmental influence. Salinity tolerance occurs even earlier in the chum salmon *Oncorhynchus keta* (Fig. 5.4c). However, here the onset is under greater genetic control than is found in either of the other two species. The actual mechanistic basis of salinity tolerance (i.e. increased gill 'chloride cells', gill Na^+/K^+-ATPase activity, membrane permeability and drinking rate), is roughly the same in each case; it is only the timing and control of onset which differs. Unlike the example of crustacean heart function, the physiological itineraries for two of these fish

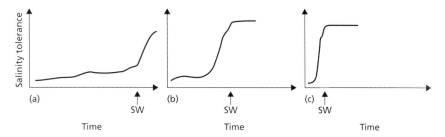

Fig. 5.4 Comparison of the timing of the onset of salinity tolerance in three species of anadromous salmonids: (a) the brook trout *Salvelinus fontinalis* (onset: ontogeny and sea water); (b) the Atlantic salmon *Salmo salar* (onset: ontogeny and photoperiod); and (c) the chum salmon *Oncorhynchus keta* (onset: ontogeny alone). SW indicates timing of exposure to sea water. (After McCormick 1994.)

species can be modified to a large degree; exactly how large and to what extent environmental modulation can account for species differences is of considerable significance.

In both birds and mammals, species exhibiting precocial development show an early development of thermoregulatory ability when compared with altricial species (Vleck *et al.* 1979; Bucher 1986; Webb and McClure 1989; Whittow and Tazawa 1991; Choi *et al.* 1993; Visser and Ricklefs 1993). However, yet again, the difference is more closely related to the timing of key physiological events than to any qualitative difference in physiology. What is interesting is the extent to which it is possible to modify the development of thermoregulation in individuals within, say, an altricial species compared with a precocial one (cf. Sect. 2.3.3). While it may be unlikely that species differences can in their entirety be accounted for by modifying the within-individual physiological itinerary, one should still be clear as to what proportion of the variation can be accounted for in this way.

When comparing the sequence and timing of events it may be useful to identify particular physiological traits that are common across a number of species or even groups of species (e.g. Derrickson 1992), so-called 'stagemarks' (Reynolds 1949). Reynolds (1949) compared the development of physiological regulations in a hamster, rat, rabbit and guinea pig by recognizing nine stagemarks, each representing the onset of a complex regulation. Although the timing was different due to different rates of development, all four species reached each stagemark in the same sequence (cf. similar results from a comparison of 147 stagemarks of prenatal structure in mouse and human; Otis and Brent 1954). What happens when higher taxonomic groupings are compared? Respiratory and cardiovascular development differs greatly both within and between the lower vertebrate classes. However, for some physiological events, prominent differences have to do with differences in timing (Burggren and Pinder 1991). For instance, while in fish the development of a

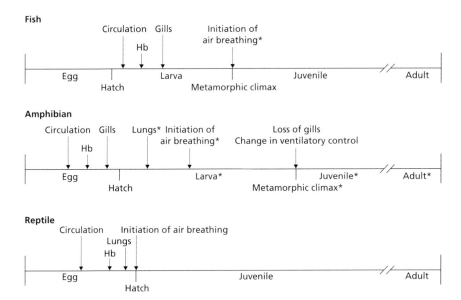

Fig. 5.5 Development of respiratory and cardiovascular function in relation to selected life history stages (*, variable) in lower vertebrates. (After Burggren and Pinder 1991.)

circulatory system, a respiratory pigment and adult respiratory exchange surfaces can often occur posthatch, their development in reptiles is compressed into the egg stage (Fig. 5.5). Similarly, while many of the physiological processes required for air-breathing occur at or before some metamorphic climax sometime after hatching in amphibians and air-breathing fish, in reptiles these processes appear before hatching so that the emergent hatchling is born with the ability to breathe air. While the taxa covered by Burggren and Pinder's (1991) analysis are at a much higher level (vertebrate classes) than those examined by Reynolds (1949) (mammal species), the principle is the same: the sequence in appearance of physiological regulations may be identical, but the timing is different. However, there may be cases where even the sequencing of events is different. Take, for example, a comparison of opossums and rats made by Adolph (1968, pp. 117–118), drawing from a number of published studies (Fig. 5.6). While some of the stagemarks occur in the same sequence, including the onset of cardiac activity, birth, breathing and suckling, some clearly do not. Thus, for the opossum the ability to defend itself against body cooling comes well before weaning, while in the rat it appears shortly afterwards.

    In conclusion, there is often marked physiological heterochrony between species. However, the extent to which the heterochrony is genetically, as opposed to environmentally, determined (i.e. a product of within-individual heterochrony) has not received the attention it deserves. Certainly, when morphological heterochrony is discussed there is often lack of clarity as to

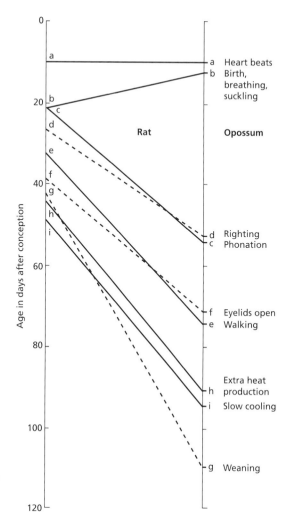

Fig. 5.6 Ages at which different activities begin in the opossum and the rat. (After Adolph 1968.)

whether between-species or within-species heterochrony is being referred to, and in the most extreme cases the two are either confused or seen as synony-mous (see the critique by Reilly *et al.* 1997). Thus, much important data are lost. What is clear from the present discussion is that an ontogenic perspective may be required not just for an understanding of within-individual variation but to fully understand the origin and meaning of between-species differences as well. In this connection it is interesting that, as long ago as the late 1960s, Adolph (1968, p. 112) observed that 'Studies on ontogeny . . . aim toward a gradual reduction in the number of items labelled as "programmed". Each one deleted from the program is added to the list of triggered (i.e. environmentally determined) items. But never will the program be empty, for no development occurs without a germ and its genetic complement'.

5.2.5 Summary

In summary, between-species variation in physiological traits can derive from (i) the different methods and protocols used to measure or compare a particular trait, and the difficulties in making meaningful comparisons between studies; (ii) differences in genetics, including species-specific differences in the timing of physiological events during development, and the extent to which species are closely or distantly related, or share similar environments; and (iii) the specific environmental circumstances that individuals of a species find themselves in. Recognizing the sources of variation is relatively straightforward, although estimating how physiological variation is apportioned between these sources is not.

5.3 Allometry

Almost synonymous, for some, with the term 'physiological diversity' are allometric relationships in which physiological traits are regressed against measures of body size (often mass). Graphs of such relationships, in which individual data points each represent a species, can be found in many textbooks and discussions of comparative animal physiology. The effect of allometry on, and implications of scaling for, between-species comparisons has been explored and discussed at great length both by physiologists (e.g. Schmidt-Nielsen 1984) and ecologists (e.g. Peters 1983; Calder 1984). Suffice to say that many (e.g. heart rate, rates of metabolism and ammonia excretion), although not all, physiological traits scale with body mass in a predictable fashion, and the mathematical formulations of these relationships are readily obtained from the literature.

However, many of these studies (i) treat each species as an independent data point; and (ii) often mix data derived from within- and between-species studies. The assumption underlying both approaches is that variation in body mass can explain nearly all of the physiological variation observed. And yet, as we shall see, there is good reason to believe that neither practice is helpful and in some cases can be completely misleading. In an early study, examining the relationship between water turnover and body mass in adult mammals, Adolph (1949) found that the allometric exponent of the relationship between species differs significantly from that within species. In comparisons between adults of 14 different species of squamate lizards, mass accounts for 88% of the variation in metabolic rate, with a further 8% explained by body temperature and activity (Andrews and Pough 1985). Examination of residuals reveals that while familial relationships explain 16% of the variation, 45% is accounted for by assigning species to one of four ecological categories (predators [day-active, reclusive and fossorial] and herbivores). McNab (1992) found that while body mass is consistently the most important factor influencing metabolic rate in mammals, the number of other contributing

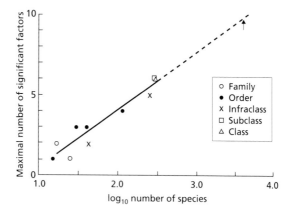

Fig. 5.7 The maximal number of significant factors correlated with basal rate of metabolism in various groups of mammals as a function of species number. The relationship is extrapolated to the number of extant mammals, ≈4000, marked by an arrow. (After McNab 1992.)

factors (e.g. phylogeny, activity, reproduction and feeding) increases with the number of species analysed (Fig. 5.7). This results not just from an increase in sample size but also from an increase in the ecological diversity of the assemblage. McNab (1992) concludes with the observation that extrapolating using the relationship between the number of species and the number of significant factors (Fig. 5.7) about 10 contributing factors would have a significant effect on metabolism, if data were available for the 4000 or so extant mammal species (assuming, of course, the relationship remains linear).

One of the most serious problems that arises when examining between-species allometry in physiological traits is that of confusing different hierarchical levels. One important case study should illustrate this point, while also highlighting a related problem, that of the historical haze whereby original works are forgotten and/or misquoted and a particular fact, idea or concept is promulgated by someone else whose view(s) then takes pre-eminence. Rubner (1883) originally formulated what is referred to as the surface rule of energy metabolism, i.e. metabolism scales to surface area$^{2/3}$. His data were derived from a within-species comparison of oxygen uptake by different sized dogs. However, Kleiber (1932, 1947, 1961) in his work illustrated the surface rule not using Rubner's within-species data, but using between-species data for mammals derived from Voit (1901). He proceeded to show that the surface rule did not hold and replaced it by the 'mass$^{3/4}$ rule', which subsequent physiologists applied indiscriminately to either within- or between-species studies. Heusner (1982), criticizing Kleiber's mass$^{3/4}$ rule, suggested that Kleiber's search for a mass-independent metabolic rate has led to an equation where there is no room to express the metabolic effect of structural and functional differences in mammals of different sizes. He reinstated Rubner's surface rule for studies of adults of the same species and dismissed Kleiber's rule as a statistical artefact. Wieser (1984) in a 'reply' to Heusner's (1982) paper explains that while he does not think that one follows the other, 'there is an ontogeny of metabolism that should not be confused with the phylogeny of metabolism, and that if we keep the two apart we may find a more general expression in which there is room for both the 2/3 and 3/4-mass exponent'.

Between-species comparisons are frequently poor predictors of within-species relationships, reflecting real differences between the hierarchical levels, thus paralleling the situation observed in Chapter 2 where within-individual allometry could not always be predicted on the basis of between-individual studies. This said, there are other reasons why this might be so. For example, while for tree swallows *Tachycineta bicolor* only 35% of the variance in metabolic rate can be explained by body mass, when birds as a group are examined, the figure is nearer 95% (Daan *et al.* 1990). However, the low proportion of the variance explained by body mass in the case of the swallows may be due more to the fact that only a narrow range of within-species body sizes are available; the heaviest is 1.3 times the lightest compared with the 200 times when birds as a group are examined.

The possibility that between-individual variation in metabolic rate may be related to between-individual differences in the mass of particular organs or organ systems has already been discussed (Sect. 3.5.2). Such a relationship has also been proposed at, and again sometimes confused with, the between-species level. For example, Kersten and Piersma (1987) speculate that interspecific differences in basal metabolic rate (BMR) for birds reflect differences in the size of the 'metabolic machinery' of a species. Daan *et al.* (1990) subsequently demonstrated that birds with higher BMRs for their body size than would have been predicted on an allometric basis have relatively large masses of hearts and kidneys. For the 22 species which they examine, these two organs, which comprise only 0.61% of body mass, explain 50% of the variation in BMR. Similarly, island species of bird have lower BMR than mainland species of the same body size, with the explanation found in the observation that there is a positive correlation between BMR and pectoral muscle mass in many flightless birds (McNab 1994).

5.4 Spatial patterns in between-species variation

The patterns of environmental variation associated with changes in the three major spatial gradients of latitude, altitude and depth described in the previous chapter (Sect. 4.4) seem just as likely to result in between-species as in between-population patterns of physiological variation.

5.4.1 Latitude

Several patterns of between-species variation in physiology with respect to latitude are of significant ecological interest. The relationships between environmental temperature and physiological variation have perhaps attracted most attention (for accounts of the detailed effects of temperature on physiology see Elliot 1981; Clarke 1983; Laudien 1986; Cossins and Bowler 1987;

Feist and White 1989; Lee and Denlinger 1991; Prosser 1991; Davenport 1992; Storey and Storey 1992, 1996; Gordon 1993; Cossins 1994; Somero 1995, 1996, 1997; Johnston and Bennett 1996; Roberts *et al.* 1997; Wood and McDonald 1997; Cossins 1998; Pörtner and Playle 1998; also the relevant sections in recent textbooks, e.g. Randall *et al.* 1997; Schmidt-Nielsen 1997). Although the view that temperature is one of the key factors in determining the geographical distributions of animals is taken as correct, much of what has been written on the subject is based on just-so stories and/or does not bear critical inspection. Many recent studies on the effects of temperature on physiological processes concentrate on the mechanistic basis of how species (cold, temperate, tropical species respond to temperature (e.g. Storey and Storey 1992, 1996; Somero 1995, 1997; Somero *et al.* 1996).

Generally speaking, any attempt at identifying large-scale spatial patterns in physiology is widely seen as grossly oversimplifying the situation (for example, the patterns are too complex to generalize, with a formidable number of possible complicatory and confounding effects, e.g. diurnal and seasonal changes, microclimate, comparability of studies, non-direct effects of temperature and phylogenetic non-independence); or as largely irrelevant (e.g. measuring thermal tolerances only tells you that an individual can live where it lives). Both positions can be directly challenged. First, spatial patterns may be too complex to generalize, but that this is so will not become apparent until the attempt to document them has at least been made. This is true notwithstanding the contents of the previous three chapters. Second, physiological data have been used to show that an individual lives where it does, and on that basis investing time and energy in such studies has come in for considerable criticism. However, the critics often fail to appreciate that such studies also address the more interesting question of where that individual *could* live; such a question may be complex to answer but it is not irrelevant.

Thermal tolerance and latitude

Terrestrial systems. The pattern of latitudinal variation in the tolerance of bird and mammal species to temperature is reasonably well established and understood. It has been known for more than 150 years that the body temperature of polar species is not greatly different from that of temperate or tropical animals (Scholander *et al.* 1950a,b,c). Neither is there any significant latitudinal pattern in metabolism (Fig. 5.8, although when the data are examined in detail this conclusion may yet prove to be a little oversimplistic). What does show pronounced variation is the rate of heat loss to the environment (conductance), this being greater for tropical and temperate species than for polar ones (Fig. 5.9). All of this is fairly well documented (Scholander *et al.* 1950a,b,c; Scholander *et al.* 1953; see general references given above), even in elementary texts (e.g. Schmidt-Nielsen 1997, pp. 254–277) and will not be expounded further here. What one would now like to know is the range of temperatures over which tropical, temperate and polar animals can maintain

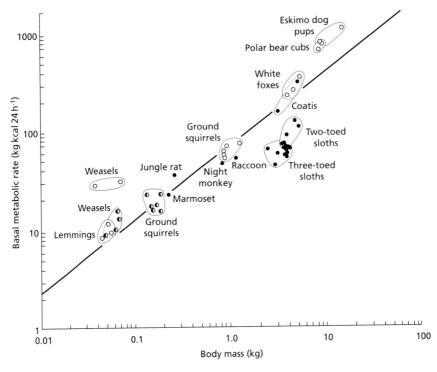

Fig. 5.8 Basal metabolic rate of Arctic (open circles), temperate (half-filled circles) and tropical (filled circles) mammals in relation to their body mass. The solid line is Kleiber's equation where metabolism is mass$^{3/4}$. (After Scholander *et al.* 1950c, with permission from the Editor of the *Biological Bulletin*.)

themselves (i.e. maintain a constant body temperature, rate of metabolism). This range is referred to as the thermoneutral zone. Anecdotal evidence would tend to suggest that the range of thermal tolerance of tropical and subtropical species can be pronounced (e.g. the televised story of a herd of elephants in a Canadian zoo breaking through ice to get a drink of water; Lindburg 1998). Presented in Figs 5.10 and 5.11 is the relationship between metabolic rate and environmental temperature for some polar (Arctic) and tropical animals (from Panama). The thermoneutral zone is recognizable in each of these graphs as the range over which there is little difference in metabolism associated with temperature. Where the thermoneutral zone ends is most obvious at lower temperatures, where there is a steep increase in metabolism at a critical temperature. This is, for instance, around 17–18°C for the Arctic weasel *Mustela rixosa*; 7–8°C for the ground squirrel *Citellus parryii*; and 27–28°C for the tropical raccoon *Procyon cacrivorus* (Fig. 5.10). As, unfortunately, the experiments were not in most cases designed to look at upper thermal tolerance it is impossible, in many cases, to extract a value for the upper limit of the thermoneutral zone. However, it is known from other studies that the variation in upper thermal limits in mammals is not nearly as pronounced as variation in lower

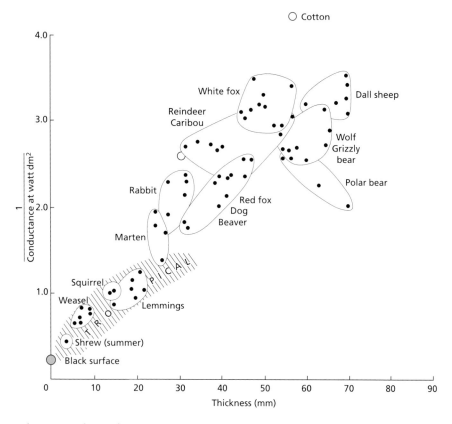

Fig. 5.9 Heat loss to the environment (conductance) in relation to fur thickness in Arctic (winter) and tropical mammals. (After Scholander *et al.* 1950b, with permission from the Editor of the *Biological Bulletin*.)

thermal limits; this is also evidenced by the fact that the data for Arctic white foxes *Alopex lagopus* and ground squirrels show no increase in metabolism even at environmental temperatures of around 30°C (Fig. 5.10).

The lower thermal tolerance of Arctic mammals is lower than that of tropical animals of equivalent size (mass), although in both sensitivity to temperature increases with decreasing animal size. If we assume that there is little variation in upper thermal tolerances between the Arctic and tropical species represented here, then the thermoneutral zone can be two to three times wider in Arctic than in tropical animals. Even the weasel, the smallest of the Arctic species examined, has a thermoneutral zone the same or perhaps even wider than that of the largest tropical species, the coati *Nasua narica*. A similar set of patterns can be seen for birds (Fig. 5.11). In short, polar mammal and bird species have a wider range of thermal tolerance than tropical species. This is notwithstanding the fact that many, particularly the small ones, make recourse to thermal refugia (Davenport 1992). Plots for laboratory mice and white rats (Fig. 5.10) show them to be animals with extremely narrow

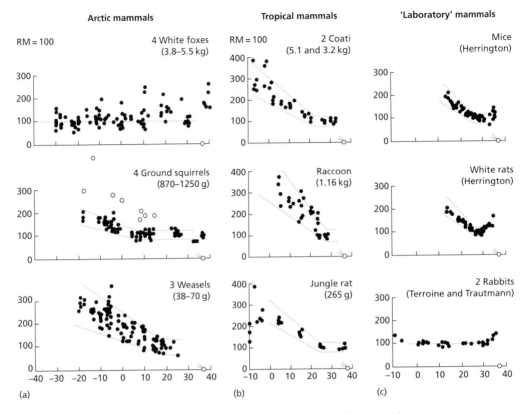

Fig. 5.10 Relationship between metabolism (expressed as a percentage of BMR) and environmental temperature for (a) Arctic, (b) tropical and (c) 'laboratory' mammal species. (After Scholander *et al.* 1950b, with permission from the Editor of the *Biological Bulletin*.)

thermoneutral zones, of just a few degrees in the case of rats. Presumably this is a result of selection in the laboratory.

Turning to ectotherms, the patterns are not as well established. Snyder and Weathers (1975), using data from Brattstrom (1968), found that after controlling for altitude the range of thermal tolerances exhibited by adult NorthAmerican amphibians increases with an increase in latitude (Fig. 5.12a). Furthermore, there is a positive relationship between the range of thermal tolerance and the environmental temperature variation encountered *in situ* (Fig. 5.12b).

Some of the most impressive ranges of thermal tolerance come not from tropical but from temperate and polar species. For example, the Antarctic mite *Ceratixodes* (= *Ixodes*) *uriae*, which can be found both free-living (normally under stones) and on the bodies of Antarctic birds, has a thermal tolerance range which extends from −30 to +40°C (Lee and Baust 1987).

The pattern of increasing thermal tolerance with increasing latitude observed in both endotherms and ectotherms (although the term 'thermal tol-

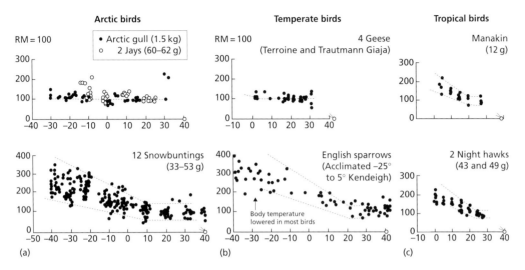

Fig. 5.11 Relationship between metabolism (expressed as a percentage of BMR) and environmental temperature for (a) Arctic, (b) temperate and (c) tropical bird species. (After Scholander *et al.* 1950b, with permission from the Editor of the *Biological Bulletin*.)

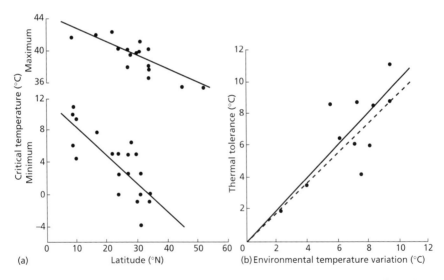

Fig. 5.12 Thermal tolerance of amphibians. (a) Relationship of upper and lower thermal tolerance (CT_{max} and CT_{min}, respectively) to latitude. (b) Relationship between range of temperature tolerance and the magnitude of environmental temperature variation. Solid line is the line of equality. (After Snyder and Weathers 1975, with permission from The University of Chicago Press.)

erance' is not directly equivalent for the two groupings) provides support for the climatic variability hypothesis, which has been used to explain Rapoport's rule (Rapoport 1982; Stevens 1989). This rule states that '[w]hen the latitudinal extent of the geographical range of organisms occurring at a given latitude is plotted against latitude, a simple positive correlation is found' (Stevens 1989). The evidence for the rule is rather poor, although studies have been strongly biased both biogeographically and taxonomically (Gaston *et al.* 1998). The climatic variability hypothesis, perhaps better termed the environmental variability hypothesis, explains Rapoport's rule as a consequence of latitudinal clines in seasonal environmental variation (Stevens 1989). For example, as the mean temperature range experienced by terrestrial animals increases with an increase in latitude, individuals require a broader range of tolerances to persist at a site, and as a consequence species can become more widely distributed.

Rapoport's rule, and hence the environmental variability hypothesis, has been argued to be associated with the general decline in species richness from low to high latitudes (Stevens 1989). More direct links between latitudinal variation in thermal tolerance and species richness have also been made. Thus, Kukal *et al.* (1991) suggest that a steep decline in the diversity of swallowtail butterfly species at high latitudes could be due to limited cold tolerance of overwintering pupae. If this is so, species with more northerly distributions should be unusually cold tolerant. They compared the northerly distributed *Papilio canadensis* with its southern relative, *P. glaucus*. Pupae were exposed for 2–5 months to four experimental treatments: outdoors in Alaska, outdoors in Michigan, constant 5°C and 25°C. Field temperatures encountered by pupae in Alaska were lower than in Michigan. The supercooling point of *P. glaucus* pupae did not vary between treatments ($= -23.5 \pm 0.52$°C [mean ± SE]). The supercooling point of *P. canadensis* pupae is not different from that of *P. glaucus* pupae, except for those kept in Alaska, when it dropped to -27.0 ± 0.55°C. Survival of pupae in Michigan was high for all populations (70–90%); in Alaska, survival of *P. canadensis* was just as high, but survival of *P. glaucus* dropped to 14%. Freezing was usually fatal in both species, but death was not immediate. No pupae survived 6 weeks at -25°C. The data support the hypothesis that winter temperatures limit swallowtail distributions.

Such large-scale spatial patterns in thermal tolerances do not seem to be obscured by the possible recourse to microclimates. None the less, this enables numbers of species with very different ranges of thermal tolerance to live in sympatry. This is illustrated by 11 species of ant that all inhabit a savannah-like grassland on the coast of north-east Spain and show a range of upper thermal tolerances (CT_{max}) from 26 to 38°C (Cerdá *et al.* 1998). Differences in thermal tolerance explains, at least in part, the temporal separation of foraging (diurnal and seasonal), allowing each of the species to exploit its own thermal niche, and so perhaps why this Spanish ant fauna is so diverse (cf. Heatwole and Muir (1989) on Tunisian ants).

Marine systems. In the shallow (<1 km) marine environment, although variation in environmental temperature increases moving away from the equator, temperate regions are the most variable (although diurnal fluctuations are small, seasonal fluctuations may be as great as 20°C). The temperature of polar waters is relatively constant at around 0°C (Cossins and Bowler 1987). For the environmental variability hypothesis to hold, one would expect both tropical and polar marine ectotherms to be characterized by a reduced temperature tolerance range when compared with equivalent temperate ectotherms.

For invertebrates, the evidence with which to test this prediction is scant, with most studies of thermal tolerance focusing on single species, and investigators working in tropical regions tending to examine upper thermal limits and those working in polar regions examining lower ones. Two studies are relevant, but conflicting. The upper lethal temperatures of five species of maldanid polychaetes correlates with their geographical ranges. For example, within the genus *Clymenella*, the upper lethal limit of the boreal-temperate zone *C. torquata* is lower than that of the warm–temperate *C. mucosa*, even after being kept at the same temperature in the laboratory (Mangum 1978). However, intertidal copepods '*Tigriopus angulatus*' from South Georgia in the South Atlantic are less tolerant of thermal extremes (particularly high temperatures) than related species, such as *T. brevicornis* from Scotland, even although the sub-Antarctic island seems to have a harsher climate (Davenport *et al.* 1997).

In marked contrast, the comparative thermal biology of fish from temperate and more particularly polar seas is the subject of an extensive literature, much of it concentrating on physiological and biochemical mechanisms of cold tolerance (Somero 1991, 1995, 1997). Broadly, polar fish species, as exemplified by the ice fish *Trematomus* sp., seem to have a greatly reduced thermal tolerance compared with more temperate species, such as *Menidia menidia* from the temperate Atlantic (Fig. 5.13). Moreover, comparing the

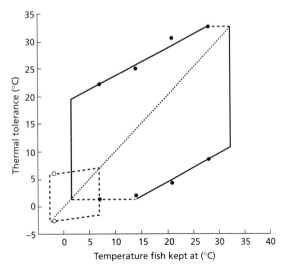

Fig. 5.13 Thermal tolerance polygons for individuals of two fish species, the temperate *Menidia menidia* (closed circles); and the Arctic *Trematomus* sp. (open circles). Broken lines represent likely relation. Dotted line indicates where lethal temperature is equal to the temperature individuals are kept at. (After Brett 1970.)

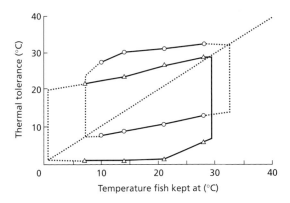

Fig. 5.14 Thermal tolerance polygons for two benthic fish species: *Pseudopleuronectes americanus* (triangles; geographical range 33–55°N) and *Spheroides maculatus* (circles; geographical range 25–45°N), both caught off New Jersey (USA) at 40°N. (Graph constructed using data presented by Brett 1970.)

thermal tolerance ranges (accounting for acclimation) of two northern hemisphere fish species, one with a more southerly distribution than the other, but using individuals caught at the same latitude, the more southerly species seems to have a smaller thermal tolerance range than the more northerly (Fig. 5.14). Ideally, one would like to compare such tolerance polygons for as many species, over as wide a range of latitudes, as possible. Unfortunately, the data that would allow this are lacking. However, the relationship between upper and lower thermal tolerances can be examined for a large number of marine fish (some studies incorporating acclimation studies, others not) across a broad latitudinal band, in a summary figure (Fig. 5.15a) put together by Brett (1970). There does, indeed, seem to be a reduction in breadth of thermal tolerance with increasing latitude, although this narrowing does not appear to be as pronounced as might have been expected from the thermal polygons for an Arctic and a temperate species (above). Clearly, the ability of the temperate fish species to acclimate and the inability of the polar fish to do so, and how one interprets these data when they are plotted, determines the magnitude of the difference between temperate and polar species. If this narrowing of tolerances is real, then, while the data uphold the environmental variability hypothesis, this finding will also have implications for Rapoport's rule itself.

Upon examining the data presented by Brett (1970), it does appear that the pattern observed for amphibians (Fig. 5.12) also holds here, namely a progressive narrowing of thermal tolerance from temperate to tropical waters (Fig. 5.15a). However, his plot is constrained by the absence of data for tropical fish (i.e. no points between 0 and 10° latitude). If data are included on the thermal tolerances of four species of tropical marine fish from Panama (Graham 1972), the pattern of narrowing of tolerances towards the equator may still be evident, but only just (Fig. 5.15b). Indeed, it is not inconceivable that from temperate to tropical latitudes upper and lower thermal tolerances are broadly parallel. If this latter scenario turned out to have some substance, it would mean that taken overall, breadth of tolerance declines as one moves away from the equator to more polar waters. A recent study of thermal tolerances of

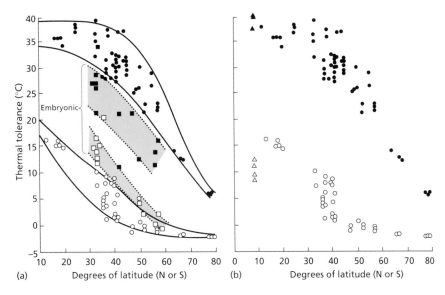

Fig. 5.15 Relationship of upper and lower thermal tolerances to latitude for marine fish. (a) Embryonic upper lethal (filled square) and lower lethal (open squares) thermal tolerances, and postembryonic upper lethal (closed circle) and lower lethal (open circle) thermal tolerance. (After Brett 1970.) (b) Data from Brett (1970) for postembryonic individuals with additional data for tropical species from Graham (1972) where upper and lower thermal tolerances are represented by filled and open triangles, respectively.

three equatorial freshwater fish species shows that while compared with more temperate species they had a greatly reduced ability to acclimate, the actual thermal polygons constructed had similar calculated areas (Fig. 5.16; but note that the data from the two regions concern different taxonomic groups of fish). These data for both marine and freshwater fish seem to run contrary to Steven's (1989) view that while there may be different climate-tolerant races within the total geographical range of a tropical species, selection for wide tolerance of single individuals would not be expected. So while there are some good data that can be marshalled to support the climatic variability hypothesis, there are also equally good data that do not. Clearly one must conclude that, with regards to describing the patterns of thermal tolerance across latitudinal gradients, the jury is still out.

Do tropical species acclimate? One of the major confounding factors in attempting to relate thermal tolerance range to latitude is acclimatization or acclimation ability. That attempting to compare thermal tolerances of individuals of different species, each with its own thermal history, is complicated is illustrated graphically in Fig. 5.15a. In constructing his figure, Brett (1970) has only included values generated by keeping individuals at temperatures within their natural range (i.e. an index of what they do rather than what they can do). Nevertheless, it appears that polar and tropical species have little acclimatory

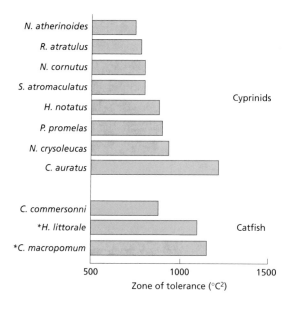

Fig. 5.16 Thermal polygon areas for a number of freshwater fish species. (All data from Brett 1956 except Amazonian fish, marked with an asterisk, from J.I. Spicer *et al.*, unpublished observation.)

ability with respect to thermal tolerance, compared with temperate species where acclimation can result in the expression of a wide range of different thermal tolerance values. But is this actually the case? The answer to this question is critical as it will determine the ways in which plots relating physiological traits to latitude should be interpreted.

Acclimatory responses to temperature are absent in tropical freshwater and marine fish (Graham 1972; J.I. Spicer, C.J. Brauner, K.J. Gaston, A. Val and C. Ballantyne, unpublished observation) and tropical lizards (Regal 1977; Feder 1978, 1982b, 1987b; Tsuji 1988). This is thought to be related to the fact that these species inhabit a thermally stable environment. Temperate species often show pronounced acclimatory ability (Sects 2.4.1, 4.2.2). However, acclimation studies are commonly only included in comparisons of particular physiological traits (which is not as often as it should be) for temperate species; often it is a matter of assumption rather than of explicit testing that polar and tropical species are incapable of acclimation. Certainly many, although not all, of the tropical anurans examined by Brattstrom (1968, 1970) show well-developed acclimatory ability. Interestingly, however, Brattstrom's (1970) study of a latitudinal gradient of anuran species in Australia seems to show that acclimatory ability is most well developed in the middle of the cline, and not at its southern end. This might be expected because mean annual temperature range and mean annual rainfall in Australia peak at intermediate latitudes (around 25°S), and decline at both higher and lower ones (Hughes *et al.* 1996). Indeed, there is also no evidence that Rapoport's rule applies in Australia (Gaston *et al.* 1998).

In summary, for both terrestrial and marine systems, there does seem to be some support for the environmental variability hypothesis, but there are also a significant number of exceptions. While Stevens' (1989) ideas on the role of environmental variability were originally framed in the context of between-species comparisons, it has already been seen that significant evidence can be marshalled for the environmental variability hypothesis from between-population studies too, although here also there do appear to be exceptions (Sect. 4.4.1, under Thermal tolerance). There is reasonably good support for the idea that the ability to acclimate is related to environmental variability; temperate animals may exhibit an overall range of thermal tolerance similar to that of comparable tropical groups, but the range present at any one time, either in summer or in winter, is likely to be reduced.

Ectotherm metabolism and latitude

While the origin of geographical differences in metabolism, and their under-pinning mechanisms, have long been the subject of controversy, the patterns were, until recently, thought to be fairly clear. Ectotherms inhabiting cold environments were characterized by greater rates of metabolism, at a given temperature, than the inhabitants of warmer environments thus, it was sug-gested, allowing the former to remain active even at low environmental tem-peratures (Ege and Krogh 1914; Krogh 1916; Bullock 1955; Prosser 1955; Vernberg 1962; Cossins and Bowler 1987). This is termed metabolic cold adap-tation (MCA). However, the concept and understanding of it have recently been the subject of close scrutiny (Clark 1980, 1991, 1993b; Johnston et al. 1998; Chown and Gaston 1999).

For terrestrial systems there is a relatively large body of literature that seems to support MCA in various invertebrates, including beetles (Aunaas et al. 1983; Strømme et al. 1986; Schultz et al. 1992; Crafford and Chown 1993; Todd 1997), grasshoppers (Chappell 1983; Massion 1983) and polar microarthropods (Block 1977; Block and Young 1978; Young 1979). There are also some relatively recent studies of a midge (Lee and Baust 1982a), a tick (Lee and Baust 1982b) and some fly species (Chown 1997), which seem to uphold the early view of Scholander et al. (1953) that MCA could not be detected in high latitude arthropod species. Admittedly, however, few of the studies have examined a large number of species along a latitudinal gradient, instead relying on comparability between disparate measurements for their conclusions.

Data generated by a number of both early and more recent studies have provided support for the presence of MCA in fish; some cold-water fish (and their isolated tissues) do seem to exhibit an elevated metabolism, compared with warm-water fish, when tested at the same temperature (Peiss and Field 1950; Scholander et al. 1953; Wohlschlag 1960, 1962; Wells 1987; Torres and Somero 1988). However, even in the early 1970s, Holeton (1974) claimed that in fact the opposite was true. Arctic fish seemed to have a reduced me-tabolism when compared with more warm-water species, a conclusion which subsequently has gained support from work by Johnston *et al.* (1991) and Steffensen *et al.* (1994). Holeton (1974) attributed the increase in metabolism in cold-water fish in earlier studies to handling stress. As it is admittedly difficult to test MCA directly using whole animals (tropical and temperate animals do not readily suffer polar temperatures and vice versa, and it can be difficult to eliminate handling stress), experiments have been carried out comparing isolated tissue respiration, mitochondrial respiration or enzyme activities from different fish assayed at different temperatures. Unfortunately, the results of these experiments have, to date, made the situation no clearer. Respiration rates of isolated brain and liver tissues from the polar cod *Boreogadus saida* (environment −1.5–2.0°C) are several times greater than those for tissues from an equivalent sized warm-water species, the golden orfe *Idus melanotus,* but only when both are measured at low (0–5°C) temperatures (Peiss and Field 1950). Examining mitochondrial activity, there is sometimes significant upregulation of mitochondrial activity in cold-water compared with warm-water fish (e.g. Crockett and Sidell 1990), but not in all (e.g. Johnston *et al.* 1994, 1998). For example, Johnston *et al.* (1998) found in studying eight perciform fish species from Antarctic, sub-Antarctic and Mediterranean waters, that differences in oxidative capacity were possibly related more to phylogeny and species-specific activity patterns than to environmental temperature *per se* (the authors note that unfortunately they did not have enough species to carry out a formal comparison). They conclude that although a number of modifications have been required to allow mitochondria to function at very low temperatures (e.g. modification of membrane lipid composition and/or enzyme binding function), the oxidative capacity of the mitochondria has not been upregulated and so is no different from similar values for warm-water species. This assumes that the mitochondrial density of warm- and cold-water forms is similar. If, however, mitochondrial density was greater in cold-water forms, then although no MCA would be detected at the subcellular level, there might still be MCA at the level of the whole organism.

Turning to studies of groups other than fish, the picture also remains confused, but more of the studies seem to support the idea that cold-water forms have a greater metabolism than warm-water ones. Fox (1936, 1938, 1939; Fox and Wingfield 1937) found that for echinoderms, crustaceans, annelids and lamellibranch molluscs, cold-water species (5–7°C, Kristineberg, Sweden) have a higher rate of oxygen uptake than warm-water species (7–15°C, Plymouth, UK). Hopkins (1946) and Scholander *et al.* (1953), working on various mollusc species, also found metabolism was greater in cold-water species compared with warm-water ones. Houlihan and Allan (1982) found that when measured at 0°C the metabolic rates of four Antarctic gastropod species were considerably greater than those measured for three temperate species. However, when both were compared at temperatures prevalent around British shores the difference disappeared, which was attributed to the fact that the Antarctic species was less susceptible to extreme cold. Whiteley *et al.* (1996) also failed to detect MCA in a comparison of a polar with a temperate isopod species. While on the basis of these studies there is reasonable support for the idea of MCA, the definitive experiments, involving a large number of species over a pronounced latitudinal cline and with appropriate controls for acclimatory differences, remain to be done.

The relationship between metabolism and latitude for aquatic ectotherms can be obscured, and in some cases eclipsed, by an opposing effect of body size. Animals in colder waters are generally larger than those in warmer waters, and, as metabolism varies inversely with body size, the former animals can have approximately the same oxygen requirements from the environment as the latter (e.g. Mangum 1963). This observation suggests that either metabolism is a more plastic trait than body size or that the advantages of being large (whatever they might be) outweigh the possible advantages that could be derived from a thermally compensated metabolic rate.

Latitude and the temperature sensitivity of metabolism in ectotherms

Given that there is some evidence that tropical species have a reduced acclimatory ability with respect to thermal tolerance (above), it might be expected that their acclimatory and acute responses of oxygen uptake (measured as an index of metabolic rate) to temperature shifts would also be reduced when compared with temperate species. One of the most complete studies of how acclimatory ability changes with latitude is that of the relationship between oxygen uptake (whole organism and isolated tissues) of a number of fiddler crab species of the genus *Uca* and the latitude at which they occur (Vernberg and Costlow 1966; Vernberg and Vernberg 1966a,b). No simple patterns emerged. For example, while oxygen uptake by isolated supraoesophageal ganglion tissue from *U. pugilator* and *U. pugnax* (both temperate species) is insensitive to incubation temperature compared with similar tissue from *U. rapax* (tropical), this pattern was reversed for a different tissue (e.g. muscle).

Tsuji (1988) compared standard metabolic rates of related lizard species, living in environments that differ in their patterns of thermal variation, to test predictions made concerning the type of acclimation response expected to be seen given the differences in the natural histories of the species examined. She predicted that the metabolism of the tropical *Sceloporus variabilis* would not show acclimation, whereas temperate winter-active *S. occidentalis* would try to compensate in order to prevent exposure to low temperatures depressing metabolism. However, she also predicted that for more northerly populations of *S. occidentalis*, where individuals are winter-dormant in order to conserve energy, there would be little attempt to compensate, and indeed metabolism would be reduced to an even greater extent than one would expect on the basis of temperature alone (what she refers to as inverse acclimation). All of these predictions were supported by the experimental data.

With regards to acute responses to temperature, there are conflicting data.

The temperature sensitivity of oxygen uptake (as a measure of metabolism) by maldanid polychaete worms is correlated with the thermal regime experienced by each species throughout its geographical range (Mangum 1963, 1972). *Clymenella torqata* occurs around a latitude of 25°N, where it experiences mean annual temperature variation in the region of 20°C. The slope of the curve relating oxygen uptake to temperature changes little within an environmentally realistic temperature range; there is a seven-fold difference in metabolism. This eurythermal species shows little evidence of any sharp reduction in temperature sensitivity being characteristic of any particular small portion of its broad temperature range. In the more southerly *C. mucosa* (16°N), which experiences mean annual temperature variation in the region of 6°C, there are pronounced changes in the slope of the oxygen uptake–temperature relationship with evidence of some regions of thermal independence. If this species was exposed to annual temperature changes of the magnitude experienced by *C. torquata* it would be subject to a 35-fold variation in oxygen uptake, but as it is, over the narrower temperature range it does experience there is only about five-fold variation.

In the case of these aquatic polychaetes, thermal sensitivity of metabolism is greater in the more southerly species. However, this contrasts markedly with the finding that thermal sensitivity is greater in high- when compared with low-latitude lizard species (Patterson 1984). Rao and Bullock (1954), using data derived from Scholander *et al.* (1953), show that at any given experimental temperature the thermal sensitivity of tropical species of crustaceans is greater (i.e. has a higher Q_{10} value) than for Arctic species (Fig. 5.17). They also note that the difference is still marked when temperatures that give equivalent metabolism (approximately 5–10°C difference, below 15°C) are compared, although there is no difference if probable habitat temperatures (Arctic 0–5 and 25–30°C) are used.

In conclusion, there is some evidence that the metabolism of tropical species shows very little capacity for acclimation, and is also more sensitive to acute temperature changes when compared with temperate species, but taken as a whole, the data are both sketchy and equivocal. Instead, a more targeted approach to the question of where acclimation responses would be expected is required, formulating predictions firmly rooted in what is known of the natural history of the species of interest (e.g. see Tsuzi 1988).

5.4.2 Altitude

Testing the environmental variability hypothesis with respect to latitudinal patterns is extremely difficult given that one must travel literally hundreds of kilometres in order to sample individuals that are experiencing different gross conditions. However, temperature differences and variation which occur over such distances can often be found within a few hundreds of metres on an altitudinal gradient. As with latitude, one might therefore expect there to be alti-

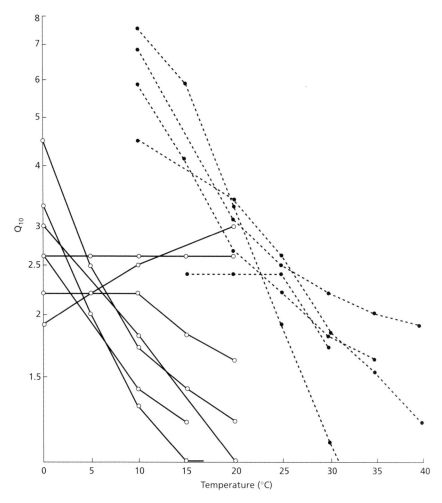

Fig. 5.17 Thermal sensitivity (measured as Q_{10}) of different species of Arctic (open circles) and tropical (closed circles) crustacean species. In the original figure legend the authors note that '(s)ome species . . . are very small or eurytopic or represented by few specimens'. (After Rao and Bullock 1954 using data derived from Scholander *et al.* 1953, with permission from The University of Chicago Press.)

tudinal patterns of physiological variation associated with environmental variability. Plotting Brattstrom's (1968) altitudinal data for anurans, controlling as far as possible for latitude (but not body size or phylogeny), a relationship between the breadth of thermal tolerance and altitude could only be detected using tropical data (9–17°N). Here, there is a negative relationship between lower, but not upper, thermal maximum temperature and altitude (Fig. 5.18). This means that the range of thermal tolerance increases with an increase in altitude, much as Snyder and Weathers (1975) found for an increase in latitude using Brattstrom's (1968) data. A similar pattern of an increasing range of thermal tolerance with an increase in altitude is also docu-

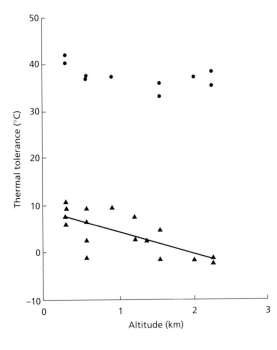

Fig. 5.18 Upper (circles) and lower (triangles) thermal tolerances for a number of anuran amphibian species from tropical latitudes along an altitudinal gradient. (Data derived from Brattstrom 1968.)

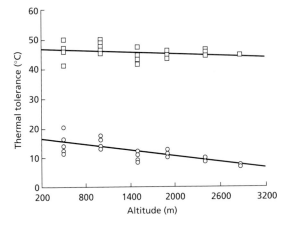

Fig. 5.19 Upper (squares) and lower (circles) thermal tolerances for a number of dung beetle species along an altitudinal gradient in South Africa. (Gaston and Chown 1999.)

mented in a recent study of dung beetle species across a transect stretching from the coast of South Africa to the Lesotho border (Fig. 5.19; Gaston and Chown 1999). In both cases, it is changes in the lower thermal tolerance (CT_{min}) that are primarily responsible for the gradient in the breadth of tolerance.

It is difficult to find other studies which have compared the thermal tolerance range of different species and related these to altitude. There appear to be three problems.

1 Many investigators examine either upper or lower thermal tolerances but

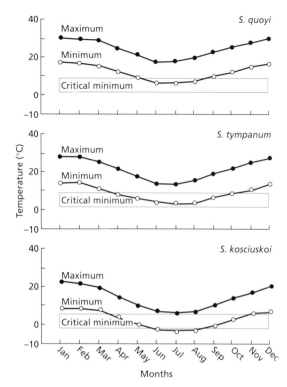

Fig. 5.20 Critical minimum temperature range for three lizard (*Spenomorphus*) species. Also shown are the mean daily maximum and minimum air temperatures in each of the three species' distribution areas. (After Spellenberg 1973, with permission from Springer-Verlag.)

seldom both. Figure 5.20 shows a comparison of lower thermal tolerances (expressed as CT_{min}, after being kept at different temperatures) of three allopatric *Sphenomorphus* lizard species, inhabiting different altitudes in south-east Australia. Also depicted are the mean daily maximum and minimum air temperatures that each of the three species actually encounter. The critical minimum ranges remained above the level of the winter daily minimum air temperature in all of the species, with the exception of *S. kosciuskoi* (Spellenberg 1973). However, nothing is known of how the upper thermal tolerance can or does change within and between these species. Similarly, species of rheophilic turbellarian triclads succeed each other in clear-cut zones along mountain streams in south-east France. *Crenobia alpina*, which lives above 800–900 m altitude has an upper incipient lethal temperature (50% mortality, indefinite time) of 12°C compared with a value of 16°C for *Polycelis felina*, which is found at 300–800 m altitude, and a value of 21°C for *Dugesia gonocephala*, which is found at altitudes below 200 m (Pattee *et al.* 1973). This difference would appear to be genetically determined, because when populations of *C. alpina* were washed downstream by floods to lower altitudes, they did not survive for long at high summer temperatures and when all three species were reared outdoors on the plains, starting in winter, *C. alpina* died in the spring, *P. felina* later and *D. gonocephala* later still. However, the investigators found it extremely difficult accurately to determine a lower thermal limit for these species, as each seemed capable of being frozen and thawed.

2 One measure of physiology can respond differently to exposure to a range of temperatures than another, even when the two should be closely related. For instance, Van Berkum (1988) found that the tolerance ranges (total ranges of temperatures over which individuals can locomote) of tropical lizards are narrower than those of temperate-zone lizards. However, there is no consistent difference between performance breadths (ranges of temperatures over which sprint speed is high) of tropical and temperate lizards.

3 Knowledge of the systematics of the species under investigation may be incomplete and/or the study may confuse different hierarchical levels of potential physiological variation. For instance, Heatwole *et al.* (1965) found a negative relationship between elevation and upper thermal tolerance (CT_{max}) for the frog '*Eleutherodactylus portoricensis*'. Subsequently, this 'species' was found to consist of two distinct species: one widespread (*Eleutherodactylus coqui*) and one restricted to high elevations (*E. portoricensis*). When CT_{max} was re-examined for these two species (Christian *et al.* 1988) it was found that differences previously attributed to elevation are, in fact, interspecific variation. There are no differences in CT_{max} of *E. coqui* from different elevations, but *E. portoricensis* has significantly lower CT_{max} than *E. coqui*.

Tentatively, it appears that in general an increase in altitude is accompanied by an increase in breadth of thermal tolerance. Knowledge of physiological mechanisms underpinning acclimation and adaptations to altitude is reasonably good, at least for a few groups (Weithe 1964; Bouverot 1985; Mani 1990; Burggren *et al.* 1991, pp. 474–478; Monge and Léon-Velarde 1991; Ward *et al.* 1995; Weber 1995; Schmidt-Nielsen 1997, pp. 209–211; West 1998).

5.4.3 Depth

The limits to the distribution of deep-sea animals has been attributed to several abiotic factors, including pressure (Somero *et al.* 1983), temperature (Carney *et al.* 1983) and the distribution of water masses (Gage 1986). As hydrostatic pressure is directly related to water depth, with an increase of 1 atmosphere (approx. 100 kPa) every 10 m (Meadows and Campbell 1988, p. 43), pressure changes with depth will have an effect on many aquatic organisms. Most experiments on tolerance to hydrostatic pressure have been carried out using common shallow-water marine species, primarily because deep-sea species are both difficult and costly to obtain and maintain. A number of these shallow-water species can tolerate a wide range of pressures (i.e. they are eurybathic) and it may not be coincidental that these same species often belong to groups that have deep-sea representatives (Schlieper 1968). Shallow-water species that cannot tolerate increasing pressure, such as the hermit crab *Eupagurus bernhardus* and the fish *Platichthys flesus* are referred to as stenobathic-barophobics, whereas species that cannot exist except in the presence of great pressure, such as the deep-sea amphipod *Paralicella capresca* (Yayanos 1978, 1981), are referred to as stenobathic-barophilic. That some species exhibit tolerances considerably greater than those they might require *in situ* (Fig. 5.21), is exemplified *par excellence* in the recorded pressure tolerance for two species of tardigrade, *Macrobiotus occidentalis* and *Echiniscus japonicus*. Both survive high-speed compression under a hydrostatic pressure of 600 000 kPa (six times the pressure of sea water at a depth of 10 km), being maintained at this pressure and high-speed decompression as well (Seki and Toyoshima 1998).

There are now a number of laboratory studies on the pressure tolerances of deep-sea animals, for example, starfish and sea urchins (Young and Tyler 1993; Young *et al.* 1996a,b), fish (Schlieper 1972) and a range of crustaceans (MacDonald *et al.* 1972; MacDonald and Teal 1975; MacDonald and Gilchrist 1978, 1980, 1982). These indicate that the pressure tolerance of deep-sea species is considerably greater than shallow-water species, although this is not always the case for the range of thermal tolerance, which is species-specific.

As well as high pressures, inhabitants of the deep sea have to cope with low temperatures, the absence of light, a scarce food supply, and, for some

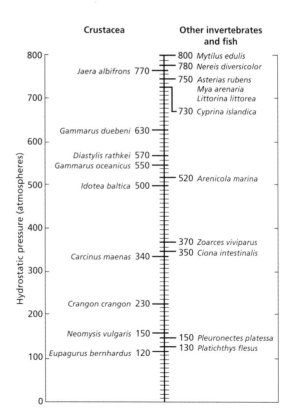

Fig. 5.21 Experimentally determined hydrostatic pressure tolerances (as LD_{50} values; 60 min at 10°C, S = 15‰) for some marine animals from shallow waters of the western Baltic Sea. (After Vernberg and Vernberg 1972, with permission from Springer-Verlag.)

deep pelagic animals, hypoxia (Vernberg and Vernberg 1972; Childress 1995; Childress and Seibel 1998).

The partial pressure of oxygen in sea water also changes with depth; there is an oxygen minimum layer over large areas of the world's oceans at depths of up to 1200 m (Meadows and Campbell 1988, p. 52). The water bodies above and below this layer, by contrast, are relatively oxygen rich. Generally speaking, the animals that inhabit the oxygen minimum layer are highly modified for extraction of oxygen from the water (via alterations in oxygen uptake and transport), and are thought to be to some extent preadapted in possessing a lower metabolic rate than surface forms (Childress and Seibel 1998).

It is a commonly held view that the rates of physiological processes are reduced in deep-living species compared with shallow-water species. Examining oxygen uptake in a number of different pelagic crustacean, cephalopod and fish species as a function of minimum depth of occurrence, there is indeed an inverse relationship in many cases (Fig. 5.22a; Childress 1971; Smith and Hessler 1974; Childress et al. 1990; Seibel et al. 1997). That this difference can be attributed to temperature differences can be discounted on the basis that such a pattern holds in Antarctic crustaceans and fish, where there is very little temperature differential between surface and deep waters (Ikeda 1988; Torres and Somero 1988). Considering some key enzymes of intermediate metabolism, not just as reliable indicators of metabolism but also as a means of controlling for the inevitable disturbance effects of capture and pressure/ temperature change on physiology, these patterns seem to hold for fish (Fig. 5.22c); measurement of such enzymes also has the advantage that species that cannot be brought back to the surface alive can still be included in analyses. In the case of pelagic cephalopods the difference in metabolism can be attributed to differences in locomotor efficiency (Seibel et al. 1997).

However, this pattern is not invariant as no such relationship between metabolism and depth has been found for pelagic chaetognaths (whole animal or isolated enzymes), cnidarians, annelid worms, copepods and molluscs (Fig. 5.22b,c; Childress and Thuesen 1993; Thuesen and Childress 1993a, 1993b, 1994; Thuesen et al. 1998). Examination of a large number of different benthic animals (ophiuroid brittle stars, holothurian sea cucumbers and burrowing and less active crustacean species) shows that there is no relationship between metabolic rate and depth that could not be accounted for by temperature differences, except in the case of those benthic crustaceans which frequently swim above and over the ocean bed.

In sum, depressed metabolic rates are not a feature of the deep-sea environment *per se* as they appear to be restricted to fairly active species of crustaceans, cephalopods and fish. The possible functional bases of these patterns are discussed by Childress (1995). He concludes that depressed metabolic rates in deep-sea species, where they are found, are not linked to the intuitively obvious candidates of food limitation and low oxygen. Instead, he suggests that high metabolic rates in shallow-water forms, with well developed vision,

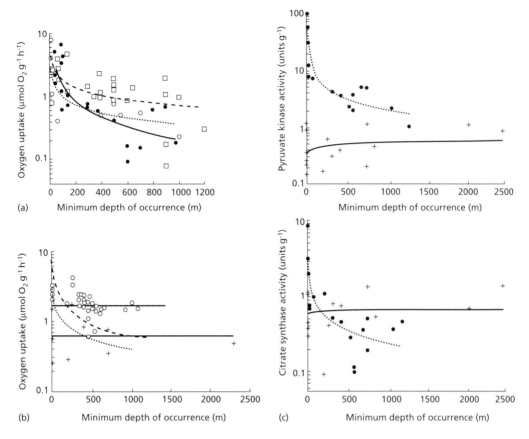

Fig. 5.22 Oxygen uptake and enzyme activities of pelagic animals as a function of minimum depths of occurrence. (a) Values for fish (open circles), crustaceans (excluding copepods) (squares) and cephalopods (closed circles). (b) Values for arrow worm (Chaetognatha) (crosses) and copepods (circles). (c) The activity of citrate synthase is related to the rate of aerobic metabolism, while pyruvate kinase is used in glycolysis. Values for fish species are represented as filled circles and values for arrow worms (Chaetognatha) are represented as crosses. (After Childress 1995, with permission from Elsevier Science; Seibel *et al.* 1997; Thuesen *et al.* 1998.)

are the result of 'selection acting to favour the use of information on predators or prey at substantial distances from an organism when ambient light is sufficient for vision at a distance'. Therefore, reduced metabolic rates result from the relaxation of this selection at greater depths where darkness prevails. Such a view presumes that detection, and reaction to, predators/prey requires less aerobic capacity in deep-sea forms than in shallow-water ones.

Childress (1995) notes that single genera, or families, can often be found throughout an entire depth range, displaying such considerable physiological (and other) variation that it often approximates to the entire variation for an assemblage; physiological divergence takes place in closely related species as a function of depth. Conversely, at a particular depth, members of different genera, families and even phyla show marked convergence in their physiological responses. In other words, physiological diversity in deep-sea animals is more likely to be a function of environment than of ancestry.

For a summary of the detailed physiological responses of deep-sea animals to temperature and pressure see Knight-Jones and Morgan (1966); Flügel (1972); Sleigh and MacDonald (1971); Hochachka and Somero (1973); MacDonald (1975); George (1981); Somero et al. (1983); Somero (1992a,b); MacDonald et al. (1993) and Childress (1995).

The definition of depth can be stretched (a little?) to include the marine intertidal zone. Here, marked variation in temperature, for example, can often be found across a gradient of only a few metres. It is thought that species living at the tops of tidal shores are more tolerant of temperature extremes than are those living closer to the low water mark (e.g. Vernberg and Vernberg 1972; Newell 1979; Underwood 1979), although this view has been questioned (McMahon 1990). Davenport and MacAlister (1996) determined the upper and lower mean lethal temperatures for all of the species that occur on the same shore at Huskvik, South Georgia: three bivalve, three gastropod, one crustacean and one annelid species. This was the first time that such a comprehensive approach had been attempted, but was made manageable by the depauperate nature of the fauna. Each of the species, with the exception of the crustacean which is restricted to tide pools, experience different degrees of emersion depending on the height at which they occur on the shore. Species living further up the shore, and so having to endure exposure to extreme air temperatures for longer—minimum and maximum temperatures of –19 and +24°C have been recorded—display an increase in thermal tolerance range (mean upper lethal temperature–mean lower lethal temperature) compared with those found lower down (Fig. 5.23). Tolerance to salinity and desiccation were also related to shore height. That such a pattern can be detected is signifi-

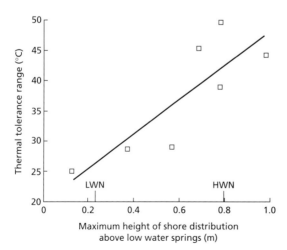

Fig. 5.23 The thermal tolerance range (upper–lower temperature tolerance) as a function of the maximum height on a rocky shore (South Georgia) for each of the seven invertebrate species present. LWN, low water neaps; HWN, high water neaps. (After Davenport and MacAlister 1996, with permission from Cambridge University Press.)

cant, as again one might think that use of microhabitat refuges (habitation of environments that buffer temperature changes, e.g. under rocks and weeds) should ameliorate potential differences. Davenport and MacAlister (1996) observe that the levels of tolerance they measured were similar to (or even less than) those exhibited elsewhere by the species examined, and suggest therefore, in line with the views of Lewis (1964) and Newell (1979), that the upper penetration of the South Georgian intertidal zone is constrained by limits of tolerance. There does, however, still seem to be a general consensus that lower limits are probably determined by biotic interactions (Connell 1961; Paine 1974).

5.5 Rare and common species

The extent to which there are systematic differences in the physiologies of rare and common species has been a topic of perhaps greater concern to ecologists than to physiologists (Lawton 1991). Much of the discussion of the issue centres on a paper by Brown (1984), who argues that locally abundant and widely distributed species tend to have broad niches, while species which are locally rare and restricted in distribution tend to have narrow niches; there tends to be a positive interspecific relationship between local abundance and regional distribution (Hanski 1982; Gaston 1996b; Gaston *et al.* 1997). Given that variation in physiology must play an important role in determining differences in the breadths of n-dimensional niches, this suggests that abundant and widespread species should have broad physiological tolerances, and rare and restricted species should have narrow tolerances. Certainly, some species do seem to have very wide environmental tolerances, which have often enabled them to become widely distributed and successfully to exploit a variety of habitats (e.g. the invasion of the coasts of the USA, Australia and South Africa by the European shore crab *Carcinus maenas*; Grosholz and Ruiz 1996). The evidence for a general relationship between abundance or range size and tolerance is, however, somewhat scattered.

van Herrewege and David (1997), investigating desiccation and starvation tolerance of 10 species of endemic and 12 species of cosmopolitan *Drosophila*, find that there is no difference between these two groupings, although there is a difference between temperate and tropical species (Fig. 5.24). On the other hand, Stanley and Parsons (1981) claim that cosmopolitan *Drosophila* species show the greatest resistance to desiccation and cold. Substantial variation in desiccation tolerance between populations of the same species may partially explain why these sets of investigators come to different conclusions (cf. Parsons and Stanley's (1981) and van Herrewege and David's (1997) interpretations of desiccation tolerance by *D. repleta*, and to a lesser extent *D. buzzatii*).

The larvae of the brine fly *Paracoenia calida* are found only in the hot (source temperature 54°C) saline waters of Wilbur Hot Springs, Colusa Co.,

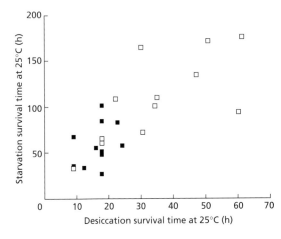

Fig. 5.24 Relationship between starvation tolerance and desiccation tolerance of temperate (open squares) and tropical (closed squares) fruit fly *Drosophila* species. (After van Herrewege and David 1997.)

California (USA). Another species of brine fly *Ephydra goedeni* can also be found in this hot spring, but unlike *P. calida*, this species has a very wide geographical range (western North America). While there was little difference in osmoregulatory capacity between the two, the restricted species was unable to tolerate anything like the range of salinities tolerated by the more widespread one (Barnby 1987). This study does seem to support the view that rare or restricted species are characterized by narrow environmental tolerances.

The catostomid fish *Moxostoma robustum* is known only from a small (and currently endangered) population which inhabits a tributary of the Altamaha River, Georgia (USA); some key physiological tolerances (to temperature, salinity, pH and hypoxia) have been determined in an attempt to evaluate the environmental requirements of this species, prior to embarking on captive breeding programmes (Walsh *et al.* 1998). This is an extremely rare fish, and yet broadly speaking its tolerances are not particularly narrow. For example, the thermal tolerance is similar to that of many other North American fish larvae/juveniles (Rombough 1996) and the range of tolerance may actually be slightly greater than that recorded for a relatively common, and more widespread, catostomid species *Catostomus commersoni* (McCormack *et al.* 1977). The extent to which the range of tolerance exhibited by *M. robustum* is a reflection of the thermal requirements of its present distribution or a relic of its past distribution (spanning south-east Atlantic coastal drainage system from North Carolina to Georgia, USA) is not known. It is clear, however, that distinction should be made between rare species which relatively recently had much greater range sizes (like *M. robustum*), and those that have, as far as can be determined, had a very limited distribution for long periods of time (like the brine fly *Ephydra goedeni*, referred to above). It may also be critical to take into account the extent to which a rare species is subject to marked local temporal

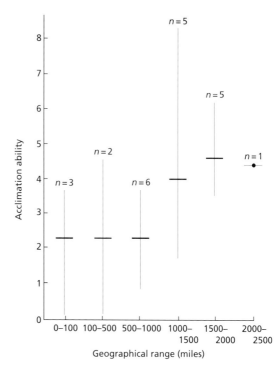

Fig. 5.25 Relationship between acclimation ability and geographical range size for Australian frog species. Acclimation ability is measured as the rate of adjustment of critical thermal maxima, for a 20°C change in the temperature at which individuals are kept. Horizontal line, mean; vertical line, range. (After Brattstrom 1970, with permission from Elsevier Science.)

environmental variability, as such variations may exceed spatial variation in environmental variability, overwhelming any patterns of spatial variation in thermal tolerance.

Brattstrom (1968, 1970) appears to have shown that, generally speaking, anuran amphibians from both Australia and the USA with restricted geographical ranges exhibit less ability to acclimate to temperature changes than do widely distributed species (Fig. 5.25). In fact, *Bufo nelsoni* (range 4 miles long, 200 yards wide, USA) and *Philoria frosti* (range three meadows on one mountain top, Australia) show little or no ability to adjust their physiologies, at least in terms of their upper thermal critical temperatures. However, there do appear to be some exceptions to this pattern, for example several wide-ranging species (at low and high latitudes) also show limited physiological plasticity and vice versa.

In conclusion, the view that rare species should have narrower tolerances than widely distributed species is an attractive one. However, formal tests remain wanting and are very difficult to perform on the basis of available data. Both here, and in many of the studies referred to in this chapter, too much reliance must be made on two species comparisons. Certainly, testing this idea should be a high priority, given the important implications for conservation, but great care must be taken in designing appropriate experiments (e.g.

breadth of tolerances of rare species should be measured for the same number of individuals as for common species).

5.6 Changing species' distributions and physiology

Concerns about possible effects of global warming have caused many eco-logical physiologists to take a renewed look at thermal relationships and the role played by temperature in establishing the distribution patterns of species and their performance capabilities (Huey and Kingsolver 1989, 1993; Huey and Bennett 1990; Hoffmann and Parsons 1991, 1997; Clarke 1993a, 1996; Dahlhoff and Somero 1993a,b; Kareiva *et al.* 1993; Fields *et al.* 1993; Holland *et al.* 1997; Wood and McDonald 1996). The role of temperature in establishing biogeographical patterning has long been recognized, but as has been seen in previous sections, the physiological and molecular mechanisms responsible for establishing and maintaining these patterns are not well resolved. Further-more, past studies in thermal biology generally have not examined effects of relatively small changes in temperature of the order predicted by many climate change models. Often the actual temperatures (or any other treatment for that matter) chosen for laboratory experiments are determined more by the desire to elicit an observable and quantifiable effect than to mimic envi-ronmental change *per se*, unless the species chosen for study inhabit an envi-ronment which experiences large (usually short-term) temperature extremes (e.g. hydrothermal vents, intertidal rock pools and polar regions). Extrapola-tion from such studies to more long-term, modest temperature change is doubtful at best. Despite a voluminous literature on the effect of temperature on physiological processes, studies are still a long way from being able to esti-mate the mimimal amounts of change in habitat temperature that are suffi-cient to elicit physiological responses resulting in altered distribution patterns of species. When one considers that the physiology of a species is not static—not only is it affected by, but can respond to, environmental change, and the extent to which this occurs is likely to be species-specific—the problem becomes particularly acute.

5.7 Conclusions and summary

'All the life of the shore . . . by the very fact of its existence there, gives evidence that it has dealt successfully with the realities of its world . . . The patterns of life as created and shaped by these realities intermingle and overlap so that the major design is exceedingly complex.' [Carson, 1965, p. 21]

Species comparisons are probably the most common expression of the in-vestigation of physiological diversity. That said, there are a number of signifi-cant difficulties in both making good comparisons and drawing inferences from such studies. These difficulties chiefly centre on the comparability of studies and problems of phylogenetic non-independence. Both can be addressed by constructing original investigations (as opposed to using data derived from the literature), employing larger numbers of species, and species for which a phylogeny is either known or can be generated (and so relatedness can be taken into account in the actual comparisons). Furthermore, when considering a 'species', it is too easy to forget that this grouping comprises not just adults, but juvenile, embryonic and senescent individuals as well (Chapter 2). What are referred to as species comparisons are almost invariably in fact adult species comparisons.

Physiological differentiation in particular functions may or may not accompany speciation. When it does, often small changes in physiology can result in (or at least accompany) quite marked and important differences in ecology. The increased access to greater altitude as a result of a few amino acid substitutions in the haemoglobin of goose species is a particularly striking example. Despite the scope for studying speciation via investigation of physiological diversity, usually morphological, ecological and behavioural studies predominate in this area. However, there is a priori no reason why this should be so, and it is entirely possible that those investigators interested in speciation *per se* may find that both the patterns and underlying mechanisms are more easily accessible by studying the generation and consequences of physiological variation at this hierarchical level. This may be particularly so in the case of between-species physiological heterochrony.

That there are spatial patterns in between-species variation is clear, even if their identities and underlying mechanisms are not always so. For birds and mammals, range of thermal tolerance increases with an increase in latitude, from tropical to polar regions. A similar increase is found for North American amphibians from tropic to temperate regions. However, the data for marine fish are more equivocal. Certainly, earlier studies seemed to point to an increase in thermal tolerance range with an increase in latitude, at least from tropical to temperate regions, but now this seems not to be the case. Although far from certain, it appears that while values for upper and lower thermal tolerances decrease with increasing latitude, the tolerance range may remain roughly constant. This concurs with what is found for freshwater fish. From temperate to polar regions, however, there does seem to be a decrease in the thermal tolerance range of marine fish.

The theory of metabolic cold adaptation seems to hold for many terrestrial and aquatic groups, although, in both cases, there is a significant number of exceptions. It is difficult to see how this discrepancy can be resolved without further experimental work. Although there are some exceptions, it is relatively safe to say that generally speaking temperate species have the ability to acclimate while tropical species do not.

With regards to altitudinal patterns, it appears that an increase in altitude is accompanied by an increase in breadth of thermal tolerance, much the same as is sometimes found for latitude. With increasing depth, there is a decrease in rates of oxygen uptake in many unrelated groups of animals. Neither pattern is unequivocal. Regardless of which spatial pattern is considered, considerable support for the environmental variability hypothesis can be marshalled. However, here too there are important exceptions which suggest that the situation is more complicated than that assumed by the theory. While some rare species do seem to have narrower tolerance ranges than common ones, this is not invariant.

At the beginning of the chapter we identified several areas of ecology for which interspecific physiological variation may be important. Even restricting consideration of species differences as we have done, it is clear that there are great gaps in knowledge at the interface of ecology and physiology. Despite the voluminous literature, both primary and secondary, dealing with comparisons of different species, new and good species comparisons are not yet redundant. However, such enterprises, if they are to beat the law of diminishing returns, must be tightly focused and carefully constructed.

Chapter 6: Overview

'Generalization means simplification. Such simplifications are common in physiology . . . Simplification, like generalization is necessary . . . The fallacies arise when we become deceived by these simplifications and when we take these as the representation of the real world.' [Florey, 1987]

Chapters 2–5 constitute a brief investigation of the origin, character and distribution of physiological diversity or physiological variation at the level of the individual, population, species and assemblage. A number of key themes or principles emerges. In this, the final chapter, those themes and principles are identified and discussed more explicitly. They concern, respectively, the pervasive and hierarchical nature of physiological variation (including difficulties and lessons in handling this variability), the basic patterns and their underlying mechanisms (both in terms of what is, and what remains to be, known) and their wider relevance. Each of these will be addressed in turn.

6.1 Pervasiveness

Physiological variation pervades each of the different hierarchical levels at which it has been examined. At no level is variation absent, or even minimal. It would appear that physiological diversity is the norm in animals, even when considering regulata which act to reduce variation to within particular limits. Few, if any, physiological traits appear to be as finely tuned as they have been portrayed (often implicitly), particularly in many older textbooks. For example, consider the marked variation in what is perhaps regarded as one of the most tightly regulated systems, namely body temperature in eutherian mammals, presented here as a between-species comparison, an integration of the variation encountered below the level of the species (Fig. 6.1).

While measurement or experimental error and developmental noise can account for some of the variation in physiological traits (Sect. 3.5.1), they are unlikely to explain it in its entirety or often even in major part. Much of the physiological variation observed does seem to be real, it exhibits pattern (Sect. 6.4) and has biological significance. One can distinguish four broad responses by investigators to the existence of physiological variation, whatever its origin.

First, it can simply be ignored. This is precisely what has been done in much of the older physiology literature, and not infrequently seemingly to good effect. As long as it is possible to discriminate, using appropriate statistical analysis, between control and experimental treatments, arguably the science is sound. But even applying statistical tests requires an investigator to make

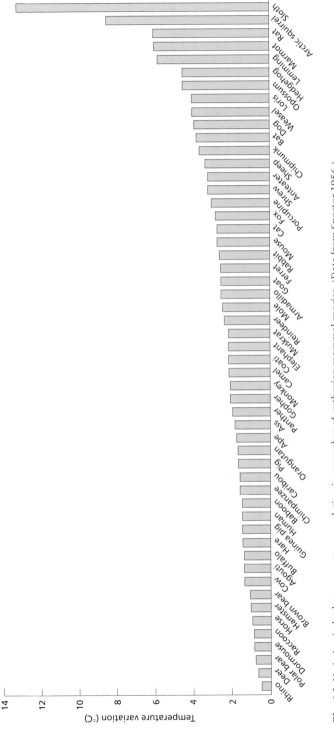

Fig. 6.1 Variation in body temperature regulation in a number of eutherian mammal species. (Data from Spector 1956.)

some assumptions about the frequency distribution of variation in the trait of interest (should parametric or non-parametric statistics be used?) and the pattern of variance among treatments (do the treatments show equal variance, which is often a prerequisite for many widely used tests?). In many cases, ignoring physiological variation at levels other than the one of interest may make little difference to the outcome of comparisons, but that does not mean that this will always, or even often, be the case.

Second, attempts can be made to minimize physiological variation. In many studies, unless this is done, treatment effects are totally swamped by the overall level of variation encountered. Minimizing variation can be achieved by, for example, using individuals which are as similar as possible (same source, maintenance conditions, body size and developmental stage). More accurate and precise equipment can be employed both for the measurement of the variables of interest and for maintaining a controlled environment for experimental individuals. Strenuous (and expensive) efforts to minimize variation in this way have been made in some quarters, such as the use of clones in ecotoxicological testing (Sect. 3.5.1). The difficulties encountered in trying to achieve such an end is convincing evidence of the pervasive nature of physiological variation. Investigators interested in minimizing this variation will seldom be content with such measures.

Third, physiological variation can be directly accounted for in studies. For example, it can be controlled for statistically, once the nature and pattern of variation has been recognized. Such an approach has always characterized the best of the ecological physiology literature. Although in such cases physiological variation may not be central to the concerns of the investigators this does not, of course, stop them from reporting in an accessible form the variability they encounter as data that may well be of vital importance to others.

Fourth and finally, the variation itself can become the object of study. A central thesis of this book is that this approach has been much neglected. And yet the fact that individuals, populations and species genuinely differ from one another in physiological traits (and within themselves at different times), and that such variation exhibits pattern, is almost certainly going to have profound implications, particularly for fields such as ecology and evolutionary biology. After all, variation is the raw material of natural selection.

The nature of physiological variation has a number of significant implications for those who study it directly.

1 At all hierarchical levels, the degree of variation in at least some physiological traits is very great. Extreme values may lie distantly from means, ranges may be very broad and variances high. Insights into such variation are likely to be severely limited if analyses are based on the small sample sizes which typify many physiological studies.

2 This variation need not even be approximately normally distributed. This may not be detectable with small sample sizes, and where it is documented will often necessitate the use of data transformations or non-parametric statistics in hypothesis testing.

3 The possible significance of extreme values and other facets of physiological variation which are not handled well by traditional statistical approaches mean that a default recourse to the use of means and variances or standard deviations may not always provide the best option for testing hypotheses concerning physiological variation (cf. Gaines and Denny 1993).

Failure to recognize these issues in the study of physiological variation could seriously limit the interpretation, and inferences, that can be drawn from published works.

6.2 Hierarchical nature

While physiological variation pervades every level of the hierarchy extending from the individual to the assemblage, it is still far from clear how this variation is distributed across that hierarchy. Just as the proportion of variance in other traits (e.g. body size) which is explained at different taxonomic levels has been explored in the context of the comparative method (e.g. Read and Harvey 1989; Harvey and Pagel 1991; Gaston 1998), it is highly desirable to understand at which of the levels of individual, population, species and assemblage most variation in physiological traits occurs. This is a significant issue both from heuristic and applied standpoints. With regard to the former, if physiological diversity is a basic fact of life, then it would seem important to understand how it is distributed. From a more practical point of view, such an understanding would give some guidance as to the circumstances in which physiological variation at levels other than that which is being explicitly studied may or may not be of major concern.

From previous chapters it should be apparent that those few studies which have sought to determine something of the distribution of physiological variation across hierarchical levels have, at the most, considered two or exceptionally three of those levels. In the main, this has been as a means of discerning the patterns of variation at the level of interest, rather than for any wider motives. Even when the partitioning of variation between two levels has been examined, there may be some degree of confusion.

One quite explicit attempt to partition variance across some of the hierarchical levels that have interested us here is the study by Heatwole *et al.* (1965) of the critical thermal maxima of two Puerto Rican species of frog. Although the relative proportions are not determined, in *Eleutherodactylus portoricensis* the greatest source of variation is attributed to within-individual variation, which is interpreted as short-term fluctuations in physiological state, and experimental error. The variance attributed to between-individual differences is quite low (but still statistically significant), which the authors suggest means that the two populations from which individuals were drawn are relatively homogenous. The difference between the two populations considered, although again statistically significant, is also small. No attempt was made formally to attribute variance at the level of species, although the partitioning of variance among levels for *E. richmondi* was similar to that found for *E. portoricensis*.

Another explicit attempt to partition variance across the hierarchical levels is the study of Chown *et al.* (in press). They examine desiccation resistance of 11 South African beetle species (all belong to the family Trogidae) and at least two populations of five of those species. Maximum tolerable water loss and rate of water loss of individual beetles is partitioned between genera (50–70%), species (20–50%), populations (1–9%) and individuals (2–18%). However, once effects of body size are accounted for, maximum tolerable water loss shows greatest variance at the level of the individual, whereas rate of water loss is greatest at the species level.

A number of difficulties will be encountered by those interested in partitioning variation across the hierarchical levels. At the level of the individual, these are principally associated with replicating the individual and discriminating experimental error from within-individual variation. These problems will often be accentuated by the practical necessity of using between-individual studies to study within-individual variation (Sect. 2.1.2).

When studying between-individual variation strictly it will be necessary to establish an analysis of repeated measures, to ensure that the magnitude of the within-individual variation does not exceed that of between-individual variation. For most of such analyses which have been performed, repeatability decreases with an increase in the time interval between measurements. Given the magnitude of within-individual variation commonly observed (Chapter 2) one would perhaps expect this to be the norm for many, if not all, traits. That physiological traits also become more variable or different the longer the time period may also result because environmental conditions display a 'reddened shift' (i.e. the longer the time period the more likely a more extreme event is to be encountered). Physiological regulations and functions are rarely stable throughout the entire lifespan of an individual.

At the level of between-population and between-species comparisons, care must be taken to ensure that when physiological measurements are made for a number of populations or species and contrasted, that (in the case of species, even after phylogeny has been controlled for) it actually is differences at the hierarchical level of interest that are detected. It is possible that differences may alternatively result from the peculiarities of the populations, or even of the individuals used in the study. It is not inconceivable, for example, that unwittingly, in comparing the effect of heavy metals on individuals of what would by normal standards be regarded as metal-sensitive and metal-tolerant species, metal-tolerant and metal-sensitive populations, respectively, of the two species could be chosen. In generalizing from the results, entirely the 'wrong' conclusions could be reached as to the effects of heavy metals on the two; this is most likely if one is not aware of physiological differentiation in populations of either species.

In the light of these observations, the determination of the hierarchical patterns of variation in a variety of physiological traits for a variety of groups of animals is strongly advocated. At present, studies in this vein which are carefully performed, address all the hierarchical levels, and exploit the best of current experimental and statistical practice, would probably serve more rapidly to advance understanding of physiological variation than any others.

In ensuring that studies are indeed addressing physiological variation at the hierarchical level at which they are purporting to do so, much would be gained by maintaining clarity in the application of terminology. For example, in isolation the term 'within-species variation' could reasonably refer to within-individual variation, between-individual within-population variation, or between-individual between-population variation. Remarkably frequently it has proven extremely difficult to discern precisely what level of the hierarchy particular studies are addressing, suggesting that great caution should be exercised in this regard (cf. the recent critique of evidence for a negative relationship between developmental instability and fitness, where Clarke (1998) notes that the comparisons made are at the population, rather than as desired by the original investigators, at the individual level). Bizarrely, some studies actually combine in a single analysis data derived from across a number of different hierarchical levels. There seems little merit in such approaches, but grounds for much confusion.

6.3 Sources

Physiological variation derives from a number of sources. The identity of these sources is well known and has been discussed at length in the previous chapters. They are measurement error, developmental noise, environmental circumstances, reversible and irreversible acclimation (or acclimatization) and genetic variation. What is not clear is how much of the physiological variation observed at a given level of the hierarchy can be attributable to each of these sources, and how consistent this pattern of attribution is between levels in the hierarchy.

Obviously, the answer to this question is inextricably linked with the proportion of variation in physiological traits which is explained at different levels in the individual to species hierarchy (Sect. 6.2). None the less, as broad generalities it is suggested as a working hypothesis that the importance of developmental noise to variation in physiological traits will tend to decline, moving from the levels of the individual to that of the species, as will the importance of environmental conditions and acclimation, while the importance of genetics will tend to increase.

Still on the subject of sources, an underlying theme of much that has been said in this book is the relationship between physiological variation as a determinant of ecology, and ecology as a determinant of physiological variation. In discussion with colleagues, we have variously heard it argued first that physiology determines ecology, and therefore that in order to understand patterns in ecology a physiological perspective is paramount, and second that ecology determines physiology, and therefore an ecological perspective is essential to understanding patterns of physiological variation (the direction of the argument is well correlated with the field of study of its proponent!). The answer to whether, for example, vertebrates colonized land because they were physiologically capable of so doing, or because they were behaviourally adapted so to do, depends on the perspective one takes. Of course, neither position is emphatically the correct one. This is a classic 'chicken-and-egg' argument. What is perhaps most significant is that it highlights the close relationship between physiology and ecology, whether one is a physiologist or an ecologist.

6.4 Patterns

Within each of the hierarchical levels, patterns in physiological variation can be identified. That is, physiological variation is not distributed randomly. Furthermore, the patterns of variation are echoed at different levels in the hierarchy. Of particular relevance to ecology, physiological variation exhibits systematic patterns of spatial and temporal variation. The patterns mentioned in this book by no means constitute an exhaustive list. The generation of such a catalogue has not been our intention, although it is suggested that a detailed listing of known patterns in physiological variation would be valuable (cf. Lawton 1996).

The available evidence for each of the patterns in physiological variation which has been documented is itself variable. In many cases, it is poor, even for patterns which are widely held as being very general. For example, it strikes us as remarkable, if not rather depressing, that controversy continues to surround the relationships between metabolism or temperature tolerance and latitude at both the between-population and between-species levels (Chapters 4 and 5). In the case of the between-species patterns this is, in major part, because the analyses to date have largely concerned the contrast solely of pairs of species (see Garland and Adolph 1994 for critique), often pairs with widely differing natural and phylogenetic histories, and these are often not comparable because issues of body size and the state of acclimation have been treated in different ways. Broader studies, embracing reasonable numbers of species across wide latitudinal spans and addressing these other issues, are desperately required to establish whether there are general underlying latitudinal trends in metabolism or tolerance. The same is true of between-species relationships between tolerance and altitude, and many others.

The existing evidence for many documented patterns also derives from studies which, while at the time they were performed, may have been admirable, employed what would presently be regarded as wholly inappropriate, if not outright misleading, techniques. Many of these patterns could usefully be revisited, as it is possible that significant fresh insights would result from so doing. The application of techniques which control for the phylogenetic non-independence of species would be an obvious case in point (see Sect. 5.2.2). In a related vein, we would observe the worrying reproduction of schematic versions of graphical figures from older works in more recent publications, without reference to the (sometimes severe) inadequacies of the underlying data. Bivariate plots based on few data points, perhaps with substantial scatter, could all too easily become established in textbook material as plots of regression lines, with no reference to the hesitancy with which the original authors drew any conclusions. In one instance, which shall remain unidentified, a set of general relationships between a physiological trait and temperature for various taxonomic groups of organisms, which have been widely figured as such in many texts, each derive from the study of often just a single animal. The original authors are explicit that this is so; the texts are not, and thus imply a generality for which there is no evidence (whether that generality indeed exists or otherwise).

There are almost certainly some, if perhaps not many, significant patterns in physiological variation which remain entirely undocumented. For example, while the pattern seems likely to exist, as far as we are aware there are no published studies of the relationship between hydrostatic pressure tolerance and depth for different populations of the same species (Sect. 4.4.3). Calling for special mention is the possible relationship between physiological diversity and rarity. Such a relationship is not only inherently interesting, but is also of considerable and immediate importance to some pressing conservation issues (as are differences in traits between rare and common species more broadly; Kunin and Gaston 1993, 1997; Gaston 1994). Yet, as has been seen, understanding is rudimentary at best (Sect. 5.5).

Feder (1987a) is largely correct when he states that there is nothing to be gained by an *ad hoc* accumulation of new studies of particular patterns simply because they have not been examined for particular species or for individuals in particular places, and so on. Instead, we suggest that what is required is a targeted programme of research that seeks to address the generality of many of the patterns (i.e. a more targeted approach to individuals, populations, species, times and places). For example, there is a requirement for data on the thermal tolerances of more tropical and polar species of marine fish, that would be comparable with much of the previous published work. The possession of such data would reveal whether or not the interspecific range of thermal tolerance decreases heading from temperate to tropical waters. There are a number of other such examples scattered throughout the previous chapters.

Finally, the essential structure has been explicitly determined for very few patterns in physiological variation, even for those patterns that are reasonably well documented (cf. Gaston and Blackburn (1999) for some interesting parallels with the field of macroecology). That is, the 'anatomy' of the patterns remains largely unknown. For example, how the underlying data result in particular trends in means and variances, how within-species statistical relationships yield between-species ones, and the details of skewness, kurtosis, intercepts, slopes and so forth. The anatomy of patterns in physiological variation seems ripe for exploration and would provide a valuable step towards determining the mechanisms generating those patterns.

6.5 Mechanisms

Almost invariably, it is easy to arrive at simple mechanistic explanations for patterns in physiological variation which make good adaptive sense. There is, however, a world of difference between the generation of such hypotheses and the explanation of observed patterns. A hypothesis is not an explanation until it has been appropriately tested and in so doing has not been falsified. Thus, as well as establishing the nature and the generality of patterns in physiological variation, systematic testing is desperately required of the mechanisms which have been proposed for patterns that are already reasonably well established. Experimental manipulations are a vital part of this agenda. Correlations on their own are useful, particularly if the pattern in question is not easily open to manipulation. However, when and where manipulation is possible, such experiments should have a high priority. For example, a concept which has been central to ecological physiology for the past 40 years, if not more, namely that acclimation can be beneficial to an individual, is only now being tested rigorously and empirically in the form of the beneficial acclimation hypothesis (Sect. 3.6). For this and many other aspects of physiological variation, the question occurs and recurs: when something is so fundamental why have so few laboratory, never mind field, tests been carried out?

6.6 Key questions

Adolph (1968) concluded his ground-breaking book *Origins of Physiological Regulations* with a substantial list of what he considered to be the key questions needing to be addressed if the field was to grow and develop. Such a practice is, it is suggested, one worth emulating. Experiments which would address the questions listed below strike us as being among those which would potentially prove very valuable in the study of physiological diversity at and between different hierarchical levels. These questions are not necessarily novel and their inclusion does not mean that some experiments have not already been conducted, merely that these were inconclusive, inappropriate, insufficient and/or have not been conducted in both laboratory and field.

Within-individual

• To what extent is it possible to manipulate the physiological itinerary of an individual animal and are there any common underlying mechanisms?
• To what extent can environmental factors in the field influence the physiological itinerary of an individual animal and which factors are most important in this regard?
• To what extent does every developmental stage have within itself the complete complement of properties required for the operation of the body?
• Is it possible to develop non-invasive techniques that will allow continuous (or at least repeated) monitoring of physiological traits that at the moment are accessible only using invasive or destructive technologies?

Between-individual

• What is the effect of environmental variation on the frequency distributions of physiological traits within populations both in the laboratory and in the field?
• At what rate is variation in a physiological trait lost when natural populations are brought into the laboratory and maintained under constant conditions?
• How frequently must extreme events occur to maintain the breadth of environmental tolerance in a population?
• What is the relative importance of physiological adjustment and selective mortality in determining temporal changes in the shapes of the frequency distributions of physiological traits within populations?
• What is the relationship between within-individual and between-individual physiological heterochrony?
• How well do within-individual patterns in physiological diversity predict between-individual patterns and vice versa?
• Is it safe to assume that genetic manipulation of a particular trait allows us to alter only that one trait?
• What form does the relationship between physiological variation and fitness take for a given trait?
• How common is irreversible acclimation?
• Is acclimatization beneficial?

Between-population

• What is the nature of the relationship (if any) between fitness and between-individual or between-population physiological variation?
• Do demographic differences accentuate or ameliorate between-population differences in physiological variation?
• Do populations at the edges and centres of geographical ranges differ in

their physiological variation and what is the contribution of acclimatization to any such differences?

• Do individuals in captivity have a narrower range of physiological variation than those in the wild and are these differences sufficient to be of conservation concern?

Between-species

• What are the relative contributions of taxon, function and environment to the distribution of acclimation?

• How well do between-population patterns in physiological variation predict between-species patterns and vice versa?

• When doesn't physiological differentiation accompany speciation?

• What is the relationship between species level rarity and physiological variation?

• What is the relationship between within-individual and between-species physiological heterochrony?

General

• How does the distribution of physiological variation among its hierarchical levels change in response to changes in environmental conditions?

• How well do altitudinal changes in physiological variation predict latitudinal changes and vice versa?

• How important is physiological differentiation between-individuals, between-populations or between-species in persistence in a changing environment (e.g. with climate change)?

• How widespread is the relationship between environmental variability and environmental tolerance, capacity and performance?

6.7 Wider relevance

As outlined at the beginning of this book (Sect. 1.3), when ecology began to emerge as a distinct discipline it was essentially synonymous with what would now be regarded as a comparative approach to ecological physiology. An understanding of the latter was regarded as central to an understanding of the determinants of patterns in the geographical distribution of animals. Despite the historical parting of the ways of the two fields, this observation would still be regarded in many quarters as axiomatic. And yet, having here examined in some detail the relationships between physiological variation and ecological patterns, it is difficult not to conclude, as others before us have done, that the relationships between geographical distributions, physiology and abiotic factors (particularly temperature) remain rather poorly understood (Futuyma 1987; Lawton 1991; Brown *et al.* 1996; Chown and Gaston 1999), despite those relationships apparently having been the subject of scientific enquiry for the best part of a century.

In part, it seems reasonable to lay blame for this circumstance on the historical legacy of the carving up of fields of study. While an understanding of geographical distributions has continued to be sought by ecologists (with a notable recent resurgence in attention to this topic), many of the tools and approaches have been developed by ecological physiologists who have had a rather different agenda. However, the circumstance seems also in part to have arisen because workers have been too readily seduced by the attractiveness of the proposition that there must be ecological consequences or implications of physiological diversity. This has been taken as read when planning experiments, without being too critical as to the exact nature of the relationship. It is suggested that it may be timely to return to many of these suppositions, to ascertain the evidence for them and to plug the gaps in understanding that inevitably emerge. We hope that this book will provide a step in that process.

6.8 In conclusion

Humans are adept at sorting and classifying, as a way of making the world about them a less confusing and more manageable place. They also manipulate their environment (or attempt to) in ways that reduce variation and make their immediate surroundings, and beyond, more hospitable. This propensity for reducing variation undeniably has influenced the way they have approached the sciences, and has certainly paid dividends. However, the lesson from many fields (e.g. study of complex systems, biodiversity and chaos theory) is that the time is now right to revisit variability and in so doing find a fresh approach to our chosen subjects.

Florey (1987), in the conclusion of his address on fads and fallacies in contemporary physiology, notes that:

'The human body, no doubt, is the prime objective of physiology, but nature is more than man and our planet is still inhabited by millions of species of animals. The diversity of life cannot be ignored by physiologists and requires explanation. It is a major task of physiologists to increase our knowledge of the diversity of animal function, of the adaptation of these so varied functions to particular environments and life styles. Such knowledge will enhance our understanding of the meaning of life on this Earth and will give us a deeper insight into our own existence.'

It is suggested that if we are to take such a charge seriously we must describe, and then understand the physiological mechanisms that underpin the distribution and abundance of animals on Earth. This should be done by adopting a more methodical, targeted and manipulative approach to physiological diversity as it occurs at and between all of its hierarchical levels.

References

Abir-Am, P.G. (1985) The philosophical background of Joseph Needham's work in chemical embryology. In: *Developmental Biology*, Vol 1955: *A Conceptual History of Modern Embryology* (ed. S.F. Gilbert), pp. 159–180. Plenum Press, New York.

Adams, C.C. (1913) *Guide to the Study of Animal Ecology*. Macmillan, New York.

Adamson, S.L. (1991) Regulation of breathing at birth. *Journal of Developmental Physiology* **15**, 45–52.

Adolph, E.F. (1949) Quantitative relations in the physiological constitutions of mammals. *Quarterly Review of Biology* **32**, 89–137.

Adolph, E.F. (1968) *Origins of Physiological Regulations*. Academic Press, New York.

Adolph, E.F. (1969) Regulations during survival without oxygen in infant mammals. *Respiration Physiology* **7**, 356–368.

Airriess, C.N., McMahon, B.R., McGaw, I.J. & Bourne, G.B. (1994) Application and *in situ* calibration of a pulsed-Doppler flowmeter for blood flow measurement in crustaceans. *Journal of the Marine Biological Association of the UK* **74**, 455–458.

Akiyama, R., Ono, H., Hoechel, J., Pearson, J.T. & Tazawa, H. (1997) Non-invasive determination of instantaneous heart rate in developing avian embryos by means of acoustocardiogram. *Medical and Biological Engineering and Computing* **35**, 323–327.

Alberch, P., Gould, S.J., Oster, G.F. & Wake, D.B. (1979) Size and shape in ontogeny and phylogeny. *Paleobiology* **5**, 296–317.

Alderdice, D.F. (1988) Osmotic and ionic regulation in teleost eggs and larvae. In: *Fish Physiology*, Vol XIA (eds W.S. Hoar & D.J. Randall), pp. 253–346. Academic Press, New York.

Aldrich, J.C. (1987) Graphical analyses of bimodal frequency-distributions with an example using *Cancer pagurus*. *Journal of Zoology* **211**, 307–319.

Aldrich, J.C. (1989) Diagnosis or elucidation —two differing uses of physiology. *Marine Behaviour and Physiology* **15**, 217–228.

Aldrich, J.C., ed. (1990) *Phenotypic Responses and Individuality in Aquatic Ectotherms*. Japaga, Ashford, Ireland.

Aldrich, J.C. (1991) The use of allometry and individuality in the examination of field data for sublethal toxicological effects. *Chemosphere* **22**, 747–767.

Aldrich, J.C. & Regnault, M. (1990) Individual variations in the response to hypoxia in *Cancer pagurus* (L.) measured at the excited rate. *Marine Behaviour and Physiology* **16**, 225–235.

Allee, W.C. (1923) The effect of temperature in limiting the geographic range of invertebrates in the Woods Hole littoral. *Ecology* **4**, 341–354.

Allee, W.C. (1926) Studies in animal aggregation: Causes and effects of bunching in land isopods. *Journal of Experimental Zoology* **45**, 255–277.

Allee, W.C., Emerson, A.E., Park, O., Park, T. & Schmidt, K.P. (1949) *Principles of Animal Ecology*. Saunders, Philadelphia.

Allen, T.F.H. & Starr, T.B. (1982) *Hierarchy: Perspectives for Ecological Complexity*. University of Chicago Press, Chicago.

Alves-Gomez, J. & Hopkins, C.D. (1997) Molecular insights into the phylogeny of mormyriform fishes and the evolution of their electric organs. *Brain, Behavior and Evolution* **49**, 324–351.

Amat, F. (1983) Zygogenetical and parthenogenetical *Artemia* in the Cádiz sea side salterns. *Marine Ecology—Progress Series* **13**, 291–293.

Anderson, M.D., Williams, J.B. & Richardson, P.R.K. (1997) Laboratory metabolism and evaporative waterloss of the Aardwolf *Proteles cristatus*. *Physiological Zoology* **70**, 464–469.

Andrewartha, H.G. & Birch, L.C. (1954) *The Distribution and Abundance of Animals*. University of Chicago Press, Chicago.

Andrews, F.B. (1925) *The Resistance of Marine Animals of Different Ages*. Publications of Puget Sound Marine Biological Station **3**, 361–363. Seattle, Washington.

Andrews, R.M. & Pough, F.H. (1985) Metabolism of squamate reptiles: allometric and ecological relationships. *Physiological Zoology* **58**, 214–231.

Anger, K. (1985) Influence of salinity on larval development of the spider crab, *Hyas araneus*, reared in the laboratory. In: *Marine Biology of Polar Regions and Effects of Stress on Marine Organisms* (eds J.S. Gray & M.E. Christiansen), pp. 463–491. John Wiley, Chichester.

Ansell, A.D., Barnett, P.R.O., Bodoy, A. & Masse, H. (1986) Upper temperature tolerances of some European molluscs. I.

Tellina fabula and *Tellina tenuis*. *Marine Biology* **58**, 33–39.

Arad, Z., Toledo, C.S. & Bernstein, M.H. (1984) Development of brain temperature regulation in the hatchling mallard duck, *Anas platyrhynchos*. *Physiological Zoology* **57**, 493–499.

Arieli, R., Arieli, M., Heth, G. & Nevo, E. (1984) Adaptive respiratory variation in 4 chromosomal species of mole rats. *Experientia* **40**, 512–514.

Arieli, R., Heth, G., Nevo, E., Zamir, Y. & Neutra, O. (1986) Adaptive heart and breathing frequencies in 4 ecologically differentiating chromosomal species of mole rats in Israel. *Experientia* **42**, 131–133.

Arking, R. (1991) *Biology of Aging: Observations and Principles*. Prentice-Hall, Englewood Cliffs, New Jersey.

Armitage, K.B. (1962) Temperature and oxygen consumption of *Orchomenella chiliensis* (Heller) (Amphipoda: Gammaroidea). *Biological Bulletin* **123**, 225–232.

Audet, D. & Fenton, M.B. (1988) Heterothermy and the use of torpor by the bat *Eptesicus fuscus* (Chiroptera: Vespertilionidae): a field study. *Physiological Zoology* **61**, 197–204.

Aunaas, T., Baust, J.G. & Zachariassen, K.E. (1983) Ecophysiological studies on arthropods from Spitsbergen. *Polar Research* **1**, 235–240.

Ayres, M.P. & Scriber, J.M. (1994) Local adaptation to regional climates in *Papilio canadensis* (Lepidoptera: Papilionidae). *Ecological Monographs* **64**, 465–482.

Baden, S.P., Loo, L.-O., Pihl, L. & Rosenberg, R. (1990) Effects of eutrophication on benthic communities, including fish, Swedish west coast. *Ambio* **19**, 113–122.

von Baer, K.E. (1828) *Über Entwicklungs-geschichte der Thiere. Beobachtung und Reflexion*. Bornträger, Königsberg.

Bale, J.S. (1991) Insects at low temperature: a predictable relationship? *Functional Ecology* **5**, 291–298.

Barbee, R.W., Perry, B.D., Re, R.N., Murgo, J.P. & Field, L.J. (1994) Hemodynamics in transgenic mice with overexpression of atrial natriuretic factor. *Circulation Research* **74**, 747–751.

Barcroft, J. (1934) *Features in the Architecture of Physiological Function*. Cambridge University Press, Cambridge.

Barker, D.J.P., ed. (1992) *Fetal and Infant Origins of Adult Disease*. British Medical Association Publications, London.

Barker, D.J.P. (1998) *Mothers, Babies and Health in Later Life*, 2nd edn. Churchill Livingstone, Edinburgh.

Barnby, M.A. (1987) Osmotic and ionic regulation of two brine fly species (Diptera: Ephydridae) from a saline hot spring. *Physiological Zoology* **60**, 327–338.

Barnea, A. & Nottebohn, F. (1994) Seasonal recruitment of hippocampal neurons in adult free-ranging black-capped chickadees. *Proceedings of the National Academy of Sciences of the USA* **91**, 11217–11221.

Barnes, H. (1967) Ecology and experimental biology. *Helgolander Wissenschaftliche Meeresuntersuchungen* **15**, 6–26.

Barnes, R.S.K. (1968) Individual variation in osmotic pressure of an ocypodid crab. *Comparative Biochemistry and Physiology* **27**, 447–450.

Bartholemew, G.A. (1958) The role of physiology in the distribution of terrestrial vertebrates. *Publications of the American Association for the Advancement of Science* **51**, 81–95.

Bartholemew, G.A. (1987) Interspecific comparison as a tool for physiological ecologists. In: *New Directions in Ecological Physiology* (eds M.E. Feder, A.F. Bennett, W.W. Burggren & R.B. Huey), pp. 11–37. Cambridge University Press, Cambridge.

Battaglia, B. (1967) Genetic aspects of benthic ecology in brackish waters. *Publications of the American Association for the Advancement of Science* **83**, 574–577.

Battaglia, B. & Bryan, G.W. (1964) Some aspects of ionic and osmotic regulation in *Tisbe* (Copepoda, Harpacticoida) in relation to polymorphism and geographical distribution. *Journal of the Marine Biological Association of the UK* **44**, 17–31.

Beadle, L.C. & Cragg, J.B. (1940) Studies on adaptation to salinity in *Gammarus* spp. I. Regulation of blood and tissues and the problem of adaptation to freshwater. *Journal of Experimental Biology* **17**, 153–163.

Beaupre, S.J. (1993) An ecological study of oxygen consumption in the mottled rock rattlesnake, *Crotalus lepidus lepidus*, and the black-tailed rattlesnake, *Crotalus molossus molossus*, from two populations. *Physiological Zoology* **66**, 437–454.

Beck, M.W. (1997) A test of the generality of the effects of shelter bottlenecks in four shore crab populations. *Ecology* **78**, 2487–2503.

de Beer, G. (1958) *Embryos and Ancestors,* 3rd edn. Clarendon Press, Oxford.

Behens-Yamada, S. (1977) Geographic range limitations of the intertidal gastropods *Littorina sitkana* and. *L. planaxis*. *Marine Biology* **39**, 61–65.

Belensky, D.A., Standaert, T.A. & Woodrum, D.E. (1979) Maturation of hypoxic ventilatory response of the newborn lamb. *Journal of Applied Physiology* **47**, 927–930.

Belkin, K.J. & Abrams, T.W. (1998) The effect

of neuropeptide FMRFamide on *Aplysia californica* siphon motoneurons involves multiple ionic currents that vary seasonally. *Journal of Experimental Biology* **201**, 2225–2234.

Bell, G.H., Emslie-Smith, D. & Paterson, C.R. (1980) *Textbook of Physiology,* 10th edn. Churchill Livingstone, Edinburgh.

Belman, B.W. & Childress, J.J. (1974) Oxygen consumption of the larvae of the lobster *Panulirus interruptus* (Randall) and the crab *Cancer productus* Randall. *Comparative Biochemistry and Physiology* **44A**, 821–828.

Bennett, A.F. (1987a) Interindividual variability: an underutilized resource. In: *New Directions in Ecological Physiology* (eds M.E. Feder, A.F. Bennett, W.W. Burggren & R.B. Huey), pp. 147–169. Cambridge University Press, Cambridge.

Bennett, A.F. (1987b) Accomplishments of ecological physiology. In: *New Directions in Ecological Physiology* (eds M.E. Feder, A.F. Bennett, W.W. Burggren & R.B. Huey), pp. 1–8. Cambridge University Press, Cambridge.

Bennett, A.F. (1997) Adaptation and the evolution of physiological characters. In: *Handbook of Physiology, Section 13: Comparative Physiology,* Vol 1 (ed. W.H. Dantzler), pp. 3–10. Oxford University Press, New York.

Bennett, A.F. & Huey, R.B. (1990) Studying the evolution of physiological performance. In: *Oxford Surveys in Evolutionary Biology,* Vol 8 (eds D. Futuyma & J. Antonovics), pp. 251–284. Oxford University Press, Oxford.

Bennett, A.F. & Lenski, R.E. (1997) Evolutionary adaptation to temperature. VI. Phenotypic acclimation and its evolution in *Escherichia coli. Evolution* **51**, 36–44.

Berman, D.I. & Zhigulskaya, Z.A. (1995) Cold-resistance of the ants of the north-west and north-east of the Palaearctic region. *Acta Zoologica Fennica* **199**, 73–80.

Bernheim, H.A. & Kluger, M.G. (1976) Fever and antipyresis in the lizard *Dipsosaurus dorsalis. American Journal of Physiology* **231**, 198–203.

Berrigan, D. (1997) Acclimation of metabolic rate in response to developmental temperature in *Drosophila melanogaster. Journal of Thermal Biology* **22**, 213–218.

Berrigan, D. & Partridge, L. (1997) Influence of temperature and activity on the metabolic rate of adult *Drosophila melanogaster. Comparative Biochemistry and Physiology* **118A**, 1301–1307.

Berteaux, D., Thomas, D.W., Bergeron, J.M. & Lapierre, H. (1996) Repeatability of daily field metabolic rate in female meadow voles (*Microtus pennsylvanicus*). *Functional Ecology* **10**, 751–759.

Bertness, M.D. & Gaines, S.D. (1993) Larval dispersal and local adaptation in acorn barnacles. *Evolution* **47**, 316–320.

Beukema, J.J. (1985) Zoobenthos survival during severe winters on high and low tidal flats in the Dutch Wadden Sea. In: *Marine Biology of Polar Regions and Effects of Stress on Marine Organisms* (eds J.S. Gray & M.E. Christiansen), pp. 351–361. John Wiley, Chichester.

Bevan, G. (1976) Changes in breeding bird populations of an oak-wood on Bookham Common, Surrey, over twenty-seven years. *London Naturalist* **55**, 23–42.

Bevan, R.M., Boyd, I.L., Butler, P.J., Reid, K., Woakes, A.J. & Croxall, J.P. (1997) Heart rates and abdominal temperatures of free-ranging South Georgian Shags *Phalacrocorax georgianus. Journal of Experimental Biology* **200**, 661–675.

Bisby, F.A. (1995) Characterization of biodiversity. In: *Global Biodiversity Assessment* (ed. V.H. Heywood), pp. 21–106. Cambridge University Press, Cambridge.

Bjorklund, A.T., Hokfelt, T. & Tohyama, M. (1992) *Ontogeny of Transmitters and Peptides in the CNS.* Elsevier Science, New York.

Blackburn, T.M. & Gaston, K.J. (1994) Body size distributions: patterns, mechanisms and implications. *Trends in Ecology and Evolution* **9**, 471–474.

Blake, R., ed. (1991) *Efficiency and Economy in Animal Physiology.* Cambridge University Press, Cambridge.

Blegvad, H. (1929) Mortality among marine animals of the littoral region in ice waters. *Report of the Danish Biological Station to the Board of Agriculture (Ministry of Fisheries), Copenhagen* **35**, 49–62.

Bligh, J. & Johnson, K.G. (1973) Glossary of terms for thermal physiology. *Journal of Applied Physiology* **35**, 941–961.

Block, W. (1977) Oxygen consumption of the terrestrial mite *Alaskozetes antarcticus* (Acari: Cryptostigmata). *Journal of Experimental Biology* **68**, 69–87.

Block, W. & Sømme, L. (1982) Cold hardiness of terrestrial mites at Signy Island, maritime Antarctic. *Oikos* **38**, 157–167.

Block, W. & Young, S.R. (1978) Metabolic adaptations of Antarctic terrestrial micro-arthropods. *Comparative Biochemistry and Physiology* **61A**, 363–368.

Boddy, K. & Dawes, G.S. (1975) Fetal breathing. *British Medical Bulletin* **31**, 3–7.

Boddy, K., Dawes, G.S., Fisher, R., Pinter, S. & Robinson, J.S. (1974) Foetal respiratory movements, electrocortical, and cardiovascular responses to hypoxaemia and hypercapnea in sheep. *Journal of Physiology* **243**, 599–618.

Boersma, P.D. (1986) Body temperature, torpor, and growth in chicks of fork-tailed storm-petrels (*Oceanodroma furcata*). *Physiological Zoology* **59**, 10–19.

Boorstein, S.M. & Ewald, P.W. (1987) Costs and

benefits of behavioral fever in *Melanoplus sanguinipes* infected by *Nosema acridophagus*. *Physiological Zoology* **60**, 586–595.

Borday, V., Fortin, G. & Champagnet, A. (1997) Early ontogeny of rhythm generation and control of breathing. *Respiration Physiology* **110**, 245–249.

Boutilier, R.G., Stiffler, D.F. & Toews, D.P. (1992) Exchange of respiratory gases, ions and water in amphibious and aquatic amphibians. In: *Environmental Physiology of the Amphibians* (eds M.E. Feder & W.W. Burggren), pp. 81–124. University of Chicago Press, Chicago.

Bouverot, P. (1985) *Adaptation to Altitude — Hypoxia in Vertebrates*. Springer, Berlin.

Bowes, G., Adamson, T.M., Ritchie, B.C., Wilkinson, M.H. & Maloney, J.E. (1981) Development of patterns of respiratory activity in unanaesthetized fetal sheep *in utero*. *Journal of Applied Physiology* **50**, 693–700.

Boyle, R. (1670) New pneumatical experiments about respiration. *Philosophical Transactions of the Royal Society of London* **5**, 2011–2031.

Bozinovic, F., Novoa, F.F. & Veloso, C. (1990) Seasonal changes in energy expenditure and digestive tract of *Abrothrix andinus* (Cricetidae) in the Andes range. *Physiological Zoology* **63**, 1216–1231.

Brattstrom, B.H. (1962) Thermal control of aggregation behavior in tadpoles. *Herpetologica* **18**, 38–46.

Brattstrom, B.H. (1968) Thermal acclimation in anuran amphibians as a function of latitude and altitude. *Comparative Biochemistry and Physiology* **24**, 93–111.

Brattstrom, B.H. (1970) Thermal acclimation in Australian amphibians. *Comparative Biochemistry and Physiology* **35**, 69–103.

Brett, J.R. (1956) Some principles in thermal requirements of fishes. *Quarterly Review of Biology* **31**, 75–87.

Brett, J.R. (1970) Fish: functional responses. In: *Marine Ecology*, Vol 1: *Environmental Factors*, Part 1, Chapter 3: *Temperature* (ed. O. Kinne), pp. 515–616. Wiley-Interscience, Chichester.

Brongersma-Sanders, M. (1957) Mass mortality in the sea. In: *Treatise on Marine Ecology and Paleoecology*, Vol 1: *Ecology* (ed. J.W. Hedgepeth). *Memoirs of the Geological Society of America* **67**, 941–1010.

Brönmark, C. & Edenhamn, P. (1994) Does the presence of fish affect the distribution of tree frogs (*Hyla arborea*)? *Conservation Biology* **8**, 841–845.

Brosnan, M.J. & Mullins, J.J. (1993) Transgenic animals in hypertension and cardiovascular research. *Experimental Nephrology* **1**, 3–12.

Brown, H.A. (1967) High temperature tolerance of the eggs of a desert anuran, *Scaphiopus hammondii*. *Copeia* **1967**, 365–370.

Brown, J.H. (1984) On the relationship between abundance and distribution of species. *American Naturalist* **124**, 255–279.

Brown, J.H. (1995) *Macroecology*. University of Chicago Press, Chicago.

Brown, R.P. (1996) Thermal biology of the gecko *Tarentola boettgeri*: comparisons among populations from different elevations within Gran Canaria. *Herpetologica* **52**, 396–405.

Brown, A.C. & Terwilliger, N.B. (1998) Ontogeny of hemocyanin function in the Dungeness crab *Cancer magister*: haemolymph modulation of hemocyanin oxygen-binding. *Journal of Experimental Biology* **201**, 819–826.

Brown, B.E. & Suharson, O. (1990) Damage and recovery of coral reefs affected by El Niño related seawater warming in the Thousand Islands, Indonesia. *Coral Reefs* **8**, 163–170.

Brown, C.R. & Brown, M.B. (1998) Intense natural selection on body size and wing and tail asymmetry in cliff swallows during severe weather. *Evolution* **52**, 1461–1475.

Brown, C.R. & Cameron, J.N. (1991) The relationship between specific dynamic action (SDA) and protein synthesis rates in the channel catfish. *Physiological Zoology* **64**, 298–309.

Brown, J.H. & Feldmeth, C.R. (1971) Evolution in constant and fluctuating environments: thermal tolerance of desert pupfish (*Cyprinodon*). *Evolution* **25**, 390–398.

Brown, J.H., Marquet, P.A. & Taper, M.L. (1993) Evolution of body size: consequences of an energetic definition of fitness. *American Naturalist* **142**, 573–584.

Brown, J.H., Stevens, G.C. & Kaufman, D.M. (1996) The geographic range: size, shape, boundaries, and internal structure. *Annual Review of Ecology and Systematics* **27**, 597–623.

Bryan, G.W. & Hummerstone, L.G. (1971) Adaptation of the polychaete *Nereis diversicolor* to estuarine sediments containing high concentrations of heavy metal. I. General observations and adaptation to copper. *Journal of the Marine Biological Association of the UK* **51**, 845–863.

Bryan, A.C., Mansell, A.L. & Levison, H. (1977) Development of the mechanical properties of the respiratory system. In: *Development of the Lung* (ed. W.A. Hodson), pp. 36–51. Dekker, New York.

Bryant, S.R., Thomas, C.D. & Bale, J.S. (1997) Nettle-feeding nymphalid butterflies: temperature, development and distribution. *Ecological Entomology* **22**, 390–398.

Bucher, T.L. (1986) Ratios of hatchling and adult mass-independent metabolism: a physiological index to the altricial-precocial continuum. *Respiration Physiology* **65**, 69–84.

Bucher, T.L. & Chappell, M.A. (1997) Respiratory exchange and ventilation during nocturnal torpor in hummingbirds. *Physiological Zoology* **70**, 45–52.

Bucher, T.L., Ryan, M.J. & Bartholemew, G.A. (1982) Oxygen consumption during resting, calling and nest building in the frog *Physalaemus pustulosus*. *Physiological Zoology* **55**, 10–22.

Buddington, R.K. & Diamond, J.M. (1989) Ontogenic development of intestinal nutrient transporters. *Annual Review of Physiology* **51**, 601–619.

Buddington, R.K. & Diamond, J. (1992) Ontogenetic development of nutrient transporters in cat intestine. *American Journal of Physiology* **263**, G605–G616.

Bulger, A.J. (1986) Coincident peaks in serum osmolality and heat tolerance rhythms in seawater-accumulated killifish (*Fundulus heteroclitus*). *Physiological Zoology* **59**, 169–174.

Bulger, A.J. & Tremaine, S.C. (1985) Magnitude of seasonal effects on heat tolerance in *Fundulus heteroclitus*. *Physiological Zoology* **58**, 197–204.

Bull, J.J. (1980) Sex determination in reptiles. *Quarterly Review of Biology* **55**, 4–21.

Bullock, T.H. (1955) Compensation for temperature in the metabolism and activity of poikilotherms. *Biological Reviews* **30**, 311–342.

Bullock, T.H. & Horridge, G.A. (1965) *Structure and Function in the Nervous Systems of Invertebrates* (2 volumes). Freeman, San Francisco.

Burger, J. (1991a) Effects of incubation temperature on the behavior of hatchling pine snakes: implications for reptilian distribution. *Behavioral Ecology and Sociobiology* **28**, 297–303.

Burger, J. (1991b) Response to prey chemical cues by hatchling pine snakes (*Pituophis melanoleucas*): effects of incubation temperature and experience. *Journal of Chemical Ecology* **17**, 1069–1078.

Burger, J. (1998) Effects of incubation temperature on hatchling pine snakes: implications for survival. *Behavioral Ecology and Sociobiology* **43**, 11–18.

Burggren, W.W. (1984) Transition of respiratory processes during amphibian metamorphosis: from egg to adult. In: *Respiration in Embryonic Vertebrates* (ed. R.S. Seymour), pp. 31–53. Junk, The Hague.

Burggren, W.W. (1987) Invasive and noninvasive methodologies in ecological physiology: a plea for integration. In: *New Directions in Ecological Physiology* (eds M.E. Feder, A.F. Bennett, W.W. Burggren & R.B. Huey), pp. 251–274. Cambridge University Press, Cambridge.

Burggren, W.W. (1989) Does comparative respiratory physiology have a role in evolutionary biology (and vice versa)? In: *Comparative Insights into Strategies for Gas Exchange and Metabolism* (eds A.J. Woakes, M.K. Grieshaber & C.R. Bridges), pp. 1–13. Cambridge University Press, Cambridge.

Burggren, W.W. (1992) The importance of an ontogenic perspective in physiological studies. In: *Strategies of Physiological Adaptation, Respiration, Circulation and Metabolism* (eds S.C. Wood, R.E. Weber, A. Hargens & R. Millard), pp. 235–253. Dekker, New York.

Burggren, W.W. & Bemis, W.E. (1990) Studying physiological evolution: paradigms and pitfalls. In: *Evolutionary Innovations* (ed. M.H. Nitecki), pp. 191–223. University of Chicago Press, Chicago.

Burggren, W.W. & Doyle, M. (1986) The action of acetylcholine upon heart rate changes markedly with development in the bullfrog. *Journal of Experimental Zoology* **240**, 137–140.

Burggren, W.W. & Fritsche, R. (1995) Cardiovascular measurements in animals in the milligram range. *Brazilian Journal of Medical and Biological Research* **28**, 1291–1305.

Burggren, W.W. & Just, J.J. (1992) Developmental changes in physiological systems. In: *Environmental Physiology of the Amphibians* (eds M.E. Feder & W.W. Burggren), pp. 467–530. University of Chicago Press, Chicago.

Burggren, W.W., McMahon, B.R. & Powers, D. (1991) Respiratory functions of blood. In: *Comparative Animal Physiology*, 4th edn, Vol 1 (ed. C.L. Prosser), pp. 437–508. Wiley-Liss, New York.

Burggren, W.W. & Pinder, A.W. (1991) Ontogeny of cardiovascular and respiratory physiology in lower vertebrates. *Annual Review of Physiology* **53**, 107–135.

Burness, G.P., Ydenberg, R.C. & Hochachka, P.W. (1998) Interindividual variability in body composition and resting oxygen consumption rate in breeding tree swallows, *Tachycineta bicolor*. *Physiological Zoology* **71**, 247–256.

Burns, J.R. (1975) Seasonal changes in the respiration of pumpkinseed, *Lepomis gibbosus*, correlated with temperature, daylength and stages of reproductive development. *Physiological Zoology* **48**, 142–149.

Burridge, L.E. & Haya, K. (1997) Lethality of pyrethrins to larvae and postlarvae of the American lobster (*Homarus americanus*). *Ecotoxicology and Environmental Safety* **38**, 150–154.

Bursell, E. (1974) Environmental aspects. In: *The Physiology of Insecta*, Vol 2 (ed. M. Rockstein), pp. 1–43. Academic Press, New York.

Burton, J.F. (1995) *Birds and Climate Change*. Christopher Helm, London.

Burton, R.S. & Feldman, M.W. (1983)

Physiological effects of an allozyme polymorphism: glutamate-pyruvate transaminase and the response to hyperosmotic stress in the copepod *Tigriopus californicus*. *Biochemical Genetics* **21**, 239–251.

Bushnell, P.G. & Brill, R.W. (1991) Responses of swimming skipjack (*Katsuwonus pelamis*) and yellowfin tuna (*Thunnus albacares*) tunas to acute hypoxia and a model of their cardiorespiratory function. *Physiological Zoology* **64**, 787–811.

Butler, P.J. & Taylor, E.W. (1975) The effect of progressive hypoxia on respiration in the dogfish (*Scyliorhinus canicula*) at different seasonal temperatures. *Journal of Experimental Biology* **63**, 117–130.

Cahn, R.D., Kaplan, N.O., Levine, L. & Zwilling, E. (1962) Nature and development of lactic dehydrogenases. *Science* **136**, 962–969.

Calder, W.A. III (1984) *Size, Function and Life History.* Harvard University Press, Cambridge, Massachusetts.

Calder, W.A.III (1994) When do humming birds use torpor in nature? *Physiological Zoology* **67**, 1051–1076.

Calow, P. (1988) Quo vadis? *Functional Ecology* **2**, 113–114.

Calow, P. (1996) Variability: noise or information in ecotoxicology? *Environmental Toxicology and Pharmacology* **2**, 121–123.

Cane, L.S., Hilton, F.K., Gray, R.D. & Harris, B.W. (1984) Further evidence for behaviorly induced hypoxic conditions in subordinate mice. *Comparative Biochemistry and Physiology* **79A**, 695–699.

Cannon, R.J.C. (1987) Effects of low-temperature acclimation on the survival and cold tolerance of an Antarctic mite. *Journal of Insect Physiology* **33**, 509–522.

Carney, R.S., Haedrich, R.L. & Rowe, G.T. (1983) Zonation of fauna in the deep-sea. In: *The Sea, Ideas and Observations on Progress in the Study of the Seas*, Vol 8: *Deep-Sea Biology* (ed. G.T. Rowe), pp. 371–398. John Wiley, New York.

Carpenter, G., Gillison, A.N. & Winter, J. (1993) DOMAIN: a flexible modelling procedure for mapping potential distributions of plants and animals. *Biodiversity and Conservation* **2**, 667–680.

Carrascal, L.M., Bautista, L.M. & Lázaro, E. (1993) Geographical variation in the density of the white stork *Ciconia ciconia* in Spain: influence of habitat structure and climate. *Biological Conservation* **65**, 83–87.

Carson, R. (1965) *The Edge of the Sea.* Panther Books, London.

Case, R.M., ed. (1985) *Variations in Human Physiology.* Manchester University Press, Manchester.

Case, T.J. (1996) Global patterns in the establishment and distribution of exotic birds. *Biological Conservation* **78**, 69–96.

Cawthorne, R.A. & Marchant, J.H. (1980) The effects of the 1978/79 winter on British bird populations. *Bird Study* **27**, 163–172.

Cerdá, X., Retana, J. & Cros, S. (1998) Critical thermal limits in Mediterranean ant species: trade-off between mortality risk and foraging performance. *Functional Ecology* **12**, 45–55.

Chan, H.M., Bjerregaard, P., Rainbow, P.S. & Depledge, M.H. (1992) Uptake of zinc and cadmium by two populations of shore crabs *Carcinus maenas* at different salinities. *Marine Ecology—Progress Series* **86**, 91–97.

Chappell, M.A. (1983) Metabolism and thermoregulation in desert and montane grasshoppers. *Oecologia* **56**, 126–131.

Chappell, M.A. & Bachman, G.C. (1995) Aerobic performance in Belding's ground squirrels (*Spermophilus beldingi*): variance, ontogeny and the aerobic capacity model of endothermy. *Physiological Zoology* **68**, 421–442.

Chappell, M.A., Bachman, G.C. & Odell, J.P. (1995) Repeatability of maximal aerobic performance in Belding's ground squirrels, *Spermophilus beldingi*. *Functional Ecology* **9**, 498–504.

Chappell, M.A., Zuk, M. & Johnsen, T.S. (1996) Repeatability of aerobic performance in Red Junglefowl: effects of ontogeny and nematode infection. *Functional Ecology* **10**, 578–585.

Charmantier, G. & Charmantier-Daures, M. (1994) Ontogeny of osmoregulation and salinity tolerance in the isopod crustacean *Sphaeroma serratum*. *Marine Ecology—Progress Series* **114**, 93–102.

Charmantier, G., Charmantier-Daures, M., Bouaricha, N., Thuet, P., Aiken, D.E. & Trilles, J.-P. (1988) Ontogeny of osmoregulation and salinity tolerance in two decapod crustaceans: *Homarus americanus* and *Penaeus japonicus*. *Biological Bulletin* **175**, 102–110.

Chen, A.C. & Schleider, P.G. (1996) An analysis of excretion in the stable fly, *Stomoxys calcitrans* (L.). *Southwestern Entomologist* **21**, 43–48.

Chen, C.P., Denlinger, D.L. & Lee, R.E. Jr (1987) Cold-shock injury and rapid cold hardening in the flesh fly *Sarcophaga crassipalpis*. *Physiological Zoology* **60**, 297–304.

Childress, J.J. (1971) Respiratory rate and depth of occurrence of midwater animals. *Limnology and Oceanography* **16**, 104–106.

Childress, J.J. (1995) Are there physiological and biochemical adaptations of metabolism in deep-sea animals? *Trends in Ecology and Evolution* **10**, 30–36.

Childress, J.J., Cowles, D.L., Favuzzi, J.A. & Mickel, T.J. (1990) Metabolic rates of benthic deep-sea decapod crustaceans decline with increasing depth primarily due to the decline in temperature. *Deep-Sea Research* **37A**, 929–950.

Childress, J.J. & Seibel, B.A. (1998) Life at stable low oxygen levels: adaptations of animals to oceanic oxygen minimum layers. *Journal of Experimental Biology* **201**, 1223–1232.

Childress, J.J. & Thuesen, E.V. (1993) Effects of hydrostatic pressure on metabolic rates of six species of deep-sea gelatinous zooplankton. *Limnology and Oceanography* **38**, 665–670.

Chin, T.S. & Chen, J.C. (1988) Acute toxicity of ammonia to larvae of the tiger prawn *Penaeus monodon*. *Aquaculture* **66**, 247–253.

Choi, I.H., Ricklefs, R.E. & Shea, R.E. (1993) Skeletal muscle growth, enzyme activities, and the development of thermogenesis: a comparison between altricial and precocial birds. *Physiological Zoology* **66**, 455–473.

Chown, S.L. (1997) Thermal sensitivity of oxygen uptake of Diptera from sub-Antarctic South Georgia and Marion islands. *Polar Biology* **17**, 81–86.

Chown, S.L. & Gaston, K.J. (1999) Exploring links between physiology and ecology at macro-scales: the role of respiratory metabolism in insects. *Biological Reviews* **74**, 87–120.

Chown, S.L., Le Lagadec, M.D. & Scholtz, C.H. (in press) Partitioning variance in a physiological variable: desiccation resistance in keratin beetles (Coleoptera, Trogidae). *Functional Ecology*.

Chown, S.L., Van Der Merwe, M. & Smith, V.R. (1997) The influence of habitat and altitude on oxygen uptake in sub-Antarctic weevils. *Physiological Zoology* **70**, 116–124.

Christian, K.A. & Morton, S.R. (1992) Extreme thermophilia in a central Australian ant, *Melophorus bagoti*. *Physiological Zoology* **65**, 885–905.

Christian, K.A., Nunez, F., Clos, L. & Diaz, L. (1988) Thermal relations of some tropical frogs along an altitudinal gradient. *Biotropica* **20**, 236–239.

Churchill, T.A. & Storey, K.B. (1989) Intermediary energy metabolism during dormancy and anoxia in the land snail *Otala lactea*. *Physiological Zoology* **62**, 1015–1030.

Clarke, A. (1980) A reappraisal of the concept of metabolic cold adaptation in polar marine invertebrates. *Biological Journal of the Linnaean Society* **14**, 77–92.

Clarke, A. (1983) Life in cold water: the physiological ecology of polar marine ectotherms. *Oceanography and Marine Biology: An Annual Review* **21**, 341–453.

Clarke, A. (1991) What is cold adaptation and how should we measure it? *American Zoologist* **31**, 81–92.

Clarke, A. (1993a) Temperature and extinction in the sea: a physiologist's view. *Paleobiology* **19**, 499–518.

Clarke, A. (1993b) Seasonal acclimatization and latitudinal compensation in metabolism: do they exist? *Functional Ecology* **7**, 139–149.

Clarke, A. (1996) The influence of climate change on the distribution and evolution of organisms. In: *Animals and Temperature. Phenotypic and Evolutionary Adaptation* (eds I.A. Johnston & A.F. Bennett), pp. 377–407. Cambridge University Press, Cambridge.

Clarke, G.M. (1998) Developmental stability and fitness: the evidence is not quite so clear. *American Naturalist* **152**, 762–766.

Clark, A.G. & Wang, L. (1994) Comparative evolutionary analysis of metabolism in nine *Drosophila* species. *Evolution* **48**, 1230–1243.

Clark, H. & Fischer, D. (1957) A reconsideration of nitrogen excretion by the chick embryo. *Journal of Experimental Zoology* **136**, 1–15.

Clegg, J.S., Drinkwater, L.E. & Sorgeloos, P. (1996) The metabolic status of diapause embryos of *Artemia franciscana*. *Physiological Zoology* **69**, 49–66.

Cloudsley-Thompson, J.L. (1962) Microclimates and the distribution of terrestrial arthropods. *Annual Review of Entomology* **7**, 199–222.

Cloudsley-Thompson, J.L. (1991) *Ecophysiology of Desert Arthropods and Reptiles*. Springer-Verlag, Berlin.

Coelho, J.R. & Mitton, J.B. (1988) Oxygen consumption during hovering is associated with genetic variation of enzymes in honey-bees. *Functional Ecology* **2**, 141–146.

Colburn, E.A. (1988) Factors influencing species diversity in saline waters of Death Valley, USA. *Hydrobiologia* **158**, 215–226.

Collins, K.J., Exton-Smith, A.N. & Doré, C. (1981) Urban hypothermia: preferred temperature and thermal perception in old age. *British Medical Journal* **1**, 175–185.

Commission for Thermal Physiology of the International Union of Physiological Sciences (1987) Glossary of terms for thermal physiology, 2nd edn. *Pflugers Archiv für die Gesampte Physiologie des Menchen und der Tiere* **410**, 567–587.

Connell, J.H. (1961) The influence of interspecific competition and other factors on the distribution of the barnacle, *Chthamalus stellatus*. *Ecology* **42**, 710–723.

Conrad, M. (1983) *Adaptability—the Significance of Variability from Molecule to Ecosystem*. Plenum Press, New York.

Conte, F.P. (1984) Structure and function of the crustacean larval salt gland. *International Review of Cytology* **91**, 45–106.

Contreras, C.L. (1986) Bioenergetics and distribution of fossorial *Spalacopus cyanus* (Rodentia): thermal stress, or cost of burrowing. *Physiological Zoology* **59**, 20–28.

Cook, W.C. (1924) The distribution of the pale western cutworm, *Porosagrotis orthogonia* Morr. A study in physical ecology. *Ecology* **5**, 60–69.

Cook, W.C. (1931) Notes on predicting the probable future distribution of introduced insects. *Ecology* **12**, 245–247.

Cooper, R.L. (1998) Development of sensory processes during limb regeneration in adult crayfish. *Journal of Experimental Biology* **201**, 1745–1752.

Cornell, H.V., Hawkins, B.A. & Hochberg, M.E. (1998) Towards an empirically-based theory of herbivore demography. *Ecological Entomology* **23**, 340–349.

Cornette, J.C., Pharriss, B.B. & Duncan, G.W. (1967) Lactic dehydrogenase isozymes in the ovum and embryo of the rat. *Physiologist* **10**, 146.

Cossins, A.R. (1994) *Temperature Adaptation of Biological Membranes*. Portland Press, London.

Cossins, A.R. (1998) Cryptic clues revealed. *Nature* **396**, 309–310.

Cossins, A.R. & Bowler, K. (1987) *Temperature Biology of Animals*. Chapman & Hall, New York and London.

Costlow, J.D. Jr, Brookhout, C.G. & Monroe, R. (1960) The effect of salinity and temperature on larval development of *Sesarma cinereum* (Bosc) reared in the laboratory. *Biological Bulletin* **118**, 183–202.

Cottle, W.H. & Carlson, L.D. (1954) Adaptive changes in rats exposed to cold. Caloric exchange. *American Journal of Physiology* **178**, 305–308.

Courtney, M.A.M. & Webb, J.E. (1964) The effect of the cold winter of 1962–3 on the Heligoland population of *Branchiostoma lanceolatum* (Pallas). *Helgolander Wissenschaftliche Meeresuntersuchungen* **10**, 301–312.

Cox, K.J.A. & Fetcho, J.R. (1996) Labelling blastomeres with a calcium indicator: a non-invasive method of visualizing neuronal activity in zebrafish. *Journal of Neuroscience Methods* **68**, 185–191.

Crafford, J.E. & Chown, S.L. (1993) Respiratory metabolism of sub-Antarctic insects from different habitats on Marion Island. *Polar Biology* **13**, 411–415.

Crill, W.D., Huey, R.B. & Gilchrist, G.W. (1996) Within- and between-generation effects of temperature on the morphology and physiology of *Drosophila melanogaster*. *Evolution* **50**, 1205–1218.

Crisp, D.J. (1964a) The effects of the winter 1962–63 on the British marine fauna. *Helgolander Wissenschaftliche Meeresuntersuchungen* **10**, 313–327.

Crisp, D.J. (1964b) The effects of the severe winter 1962–63 on marine life in Britain. *Journal of Animal Ecology* **33**, 165–210.

Crisp, D.J. (1964c) Racial differences between North American and European forms of *Balanus balanoides*. *Journal of the Marine Biological Association of the UK* **44**, 33–45.

Crisp, D.J. & Ritz, D.A. (1967) Changes in temperature tolerance of *Balanus balanoides* during its life cycle. *Helgolander Wissenschaftliche Meeresuntersuchungen* **15**, 98–115.

Crockett, E.L. & Sidell, B.D. (1990) Some pathways of energy metabolism are cold adapted in Antarctic fishes. *Physiological Zoology* **63**, 472–488.

Crowson, R.A. (1981) *The Biology of the Coleoptera*. Academic Press, London.

Cruikshank, S.J. & Weinberger, N.M. (1996) Evidence for the Hebbian hypothesis in experience-dependent physiological plasticity of neocortex: a critical review. *Brain Research Reviews* **22**, 191–228.

Csada, R.D., James, P.C. & Espie, R.H.M. (1996) The 'file drawer problem' of non-significant results: does it apply to biological research? *Oikos* **76**, 591–593.

Cutts, C.J., Metcalfe, N.B. & Taylor, A.C. (1998) Aggression and growth depression in juvenile Atlantic salmon: the consequences of individual variation in standard metabolic rate. *Journal of Fish Biology* **52**, 1026–1037.

Daan, S., Masman, D. & Groenewold, A. (1990) Avian basal metabolic rates: their association with body composition and energy expenditure in nature. *American Journal of Physiology* **259**, R333–R340.

Daan, S., Masman, D., Strijkstra, A. & Verhulst, S. (1989) Intraspecific allometry of basal metabolic rate: relations with body size, temperature, composition, and circadian phase in the kestrel, *Falco tinnunculus*. *Journal of Biological Rhythms* **4**, 267–283.

Dahlhoff, E. & Somero, G.N. (1993a) Kinetic and structural adaptations of cytoplasmic malate dehydrogenases of eastern Pacific abalone (genus *Haliotis*) from different thermal habitats: biochemical correlates of biogeographical patterning. *Journal of Experimental Biology* **185**, 137–150.

Dahlhoff, E. & Somero, G.N. (1993b) Effects of temperature on mitochondria from abalone (genus *Haliotis*): adaptive plasticity and its limits. *Journal of Experimental Biology* **185**, 151–168.

Dallinger, R. (1993) Strategies of metal detoxification in terrestrial invertebrates. In: *Ecotoxicology of Metals in Invertebrates* (eds P.S. Rainbow & R. Dallinger), pp. 243–289. Lewis Publishers, Boca Raton.

Dancis, J. & Schneider, H. (1975) Physiology: transfer and barrier function. In: *The Placenta and its Maternal Supply Line* (ed. P. Gruenwald), pp. 143–198. Medical and Technical Publishing, Lancaster.

Darwin, C. (1859) *The Origin of Species by Means of Natural Selection*. John Murray, London.

Davenport, C.B. (1897) *Experimental Morphology. Part first. Effect of Chemical and Physical Agents upon Protoplasm*. Macmillan, New York and London.

Davenport, J. (1983) Oxygen and the developing eggs and larvae of the lumpfish, *Cyclopterus lumpus*. *Journal of the Marine Biological Association of the UK* **63**, 633–640.

Davenport, J. (1992) *Animal Life at Low Temperature*. Chapman & Hall, London.

Davenport, J., Barnett, P.R.O. & McAllen, R.J. (1997) Environmental tolerances of three species of the harpacticoid copepod genus *Tigriopus*. *Journal of the Marine Biological Association of the UK* **77**, 3–16.

Davenport, J. & MacAlister, H. (1996) Environmental conditions and physiological tolerances of intertidal fauna in relation to shore zonation at Huskvik, South Georgia. *Journal of the Marine Biological Association of the UK* **76**, 985–1002.

Davies, P.S. (1966) Physiological ecology of *Patella*. I. The effect of body size and temperature on respiration rate. *Journal of the Marine Biological Association of the UK* **46**, 647–658.

Davies, P.S. (1967) Physiological ecology of *Patella*. II. Effect of environmental acclimation on the metabolic rate. *Journal of the Marine Biological Association of the UK* **47**, 61–74.

Davis, M.T.B. (1974) Changes in critical temperature during nymphal and adult development in the rabbit tick, *Haemaphysalis leporispalustris* (Acari: Ixodides: Ixosisae). *Journal of Experimental Biology* **60**, 85–94.

Davis, A.J., Jenkinson, L.S., Lawton, J.H., Shorrocks, B. & Wood, S. (1998) Making mistakes when predicting shifts in species range in response to global warming. *Nature* **391**, 783–786.

Dawes, G.S., Fox, H.E., Leduc, B.M., Liggins, G.C. & Richards, R.T. (1972) Respiratory movements and rapid eye movement sleep in the foetal lamb. *Journal of Physiology* **220**, 119–143.

De Lisle, R.C., Sarras, M.P. Jr, Hidalgo, J. & Andrews, G.K. (1996) Metallothionein is a component of exocrine pancreas secretion: Implications for zinc homeostasis. *American Journal of Physiology* **271**, C1103–C1110.

Deaton, L.E., Hilbish, T.J. & Koehn, R.K. (1984) Protein as a source of amino nitrogen during hyperosmotic volume regulation in the mussel *Mytilus edulis*. *Physiological Zoology* **57**, 609–619.

Degan, A.A., Kam, M., Khokhlova, I.S., Krasnov, B.R. & Barraclough, T.G. (1998) Average daily metabolic rate of rodents: habitat and dietary comparisons. *Functional Ecology* **12**, 63–73.

Deheyn, D., Mallefet, J. & Jangoux, M. (1997) Intraspecific variation of bioluminescence in a polychromatic population of *Amphipholis squamata* (Echinodermata: Ophiuroidea). *Journal of the Marine Biological Association of the UK* **77**, 1213–1222.

Démeusy, N. (1957) Respiratory metabolism of the fiddler crab *Uca pugilator* from two different latitudinal populations. *Biological Bulletin* **113**, 245–253.

Depledge, M.H. (1990) New approaches in ecotoxicology: can inter-individual physiological variability be used as a tool to investigate pollution effects? *Ambio* **19**, 251–252.

Depledge, M.H. (1994) Genotypic toxicity: Implications for individuals and populations. *Environmental Health Perspectives* **102** (Suppl. 12), 101–104.

Depledge, M.H. & Bjerregaard, P. (1990) Explaining individual variation in trace metal concentrations in selected marine invertebrates: The importance of interactions between physiological state and environmental factors. In: *Phenotypic Responses and Individuality in Aquatic Ectotherms* (ed. J.C. Aldrich), pp. 121–126. Japaga, Ireland.

Derrickson, E.M. (1989) The comparative method of Elgar and Harvey: silent ammunition for McNab. *Functional Ecology* **3**, 123–127.

Derrickson, E.M. (1992) Comparative reproductive strategies of altricial and precocial eutherian mammals. *Functional Ecology* **6**, 57–65.

Di Loreto, S. & Balestrino, M. (1997) Development of vulnerability to hypoxic damage *in vitro* hippocampal neurons. *International Journal of Developmental Neuroscience* **15**, 225–230.

Di Michele, L. & Powers, D.A. (1982) Physiological basis for swimming endurance differences between LDH-B genotypes of *Fundulus heteroclitus*. *Science* **216**, 1014–1016.

Di Michele, L., Powers, D.A. & Di Michele, J.A. (1986) Developmental and physiological consequences of genetic variation at enzyme synthesizing loci in *Fundulus heteroclitus*. *American Zoologist* **26**, 201–208.

Diamond, J. (1994) Evolutionary matching of physiological capacities to natural loads. *Physiologist* **37**, A1.

Diamond, J.M. (1992) The red flag of optimality. *Nature* **355**, 204–206.

Diamond, J.M. (1993) Quantitative design of life. *Nature* **366**, 405–406.

Diamond, J. & Hammond, K. (1992) The matches, achieved by natural selection, between biological capacities and their natural loads. *Experientia* **48**, 551–557.

Diaz, J.A., Bauwens, D. & Asensio, B. (1996) A comparative study of the relation between heating rates and ambient temperatures in lacertid lizards. *Physiological Zoology* **69**, 1359–1383.

Diaz, R.J. & Rosenberg, R. (1995) Marine benthic hypoxia: a review of its ecological effects and the behavioural responses of benthic macrofauna. *Annual Review of Oceanography and Marine Biology* **33**, 245–303.

Dickerson, R.E. & Geis, I. (1983) *Hemoglobin: Structure, Function, Evolution and Pathology*. Benjamin-Cummings, Menlo Park, California.

Dickie, L.M. & Medcof, J.C. (1963) Causes of mass mortalities of scallops (*Placopecten magellanicus*) in the southwestern Gulf of St. Lawrence. *Journal of the Fisheries Research Board (Canada)* **20**, 451–482.

Diez, J.M. & Davenport, J. (1987) Embryonic respiration in the dogfish (*Scyliorhinus canicula* L.). *Journal of the Marine Biological Association of the UK* **67**, 249–261.

Dittman, D.E. (1997) Latitudinal compensation in oyster ciliary activity. *Functional Ecology* **11**, 573–578.

Doi, K. & Kuroshima, A. (1979) Lasting effects of infantile cold experience on cold tolerance in adult rats. *Japanese Journal of Physiology* **29**, 139–150.

Dorgelo, J. (1977) Comparative ecophysiology of gammarids (Crustacea: Amphipoda) from marine, brackish- and fresh-water habitats exposed to the influence of salinity–temperature combinations. IV. Blood sodium regulation. *Netherlands Journal of Sea Research* **11**, 184–199.

Drake, J.A., Mooney, H.A., di Castri, F., Groves, R.H., Kruger, F.J., Rejmánek, M. & Williamson, M., eds. (1989) *Biological Invasions: A Global Perspective*. John Wiley, Chichester.

Duchêne, J.C. (1985) Comparative study of respiration in a North temperate and sub-Antarctic population of *Thelepus setosus* (Annelida, Polychaeta). In: *Marine Biology of Polar Regions and Effects of Stress on Marine Organisms* (eds J.S. Gray & M.E. Christiansen), pp. 247–258. John Wiley, Chichester.

Duerr, F.G. (1967) Changes in the size–metabolic rate relationship of *Lymnaea stagnalis appressa* Say by digenetic trematode parasitism. *Comparative Biochemistry and Physiology* **20**, 391–398.

Durstewitz, G. & Terwilliger, N.B. (1997) Developmental changes in hemocyanin expression in the Dungeness crab, *Cancer magister*. *Journal of Biological Chemistry* **272**, 4347–4350.

Dutenhoffer, M.S. & Swanson, D.L. (1996) Relationship of basal to summit metabolic rate in passerine birds and the aerobic capacity model for the evolution of endothermy. *Physiological Zoology* **69**, 1232–1254.

Dutrochet, E. (1837) Observations sur la *Chara flexis*: Modifications dans la circulation de cette plante sous l'influence d'un changement de température, d'une irration mécanique, de l'action des sels etc. *Comptes Rendus de l' Academie de Sciences* **5**, 775–784.

Dybern, B.I. (1967) The distribution and salinity tolerance of *Ciona intestinalis* (L.) f. *typica* with special reference to the waters around southern Scandinavia. *Ophelia* **4**, 207–226.

Dykstra, C.R. & Karasov, W.H. (1992) Changes in gut structure and function of house wrens (*Troglodytes aedon*) in response to increased energy demands. *Physiological Zoology* **65**, 422–442.

Eakin, R.E. & Fisher, J.R. (1958) Patterns of nitrogen excretion in developing chick embryos. In: *The Chemical Basis of Development* (eds W.D. McElroy & B. Glass), pp. 514–522. Johns Hopkins Press, Baltimore.

Eaton, J.G., McCormick, J.H., Goodno, B.E., O'Brien, D.G., Stefany, H.G., Hondzo, M. & Scheller, R.M. (1995) A field information-based system for estimating fish temperature tolerances. *Fisheries* **20**, 10–18.

Ebenhard, T. (1988) Introduced birds and mammals and their ecological effects. *Swedish Wildlife Research* **13**, 1–107.

Eden, G.J. & Hanson, M.A. (1987) Maturation of the respiratory response to acute hypoxia in the newborn rat. *Journal of Physiology* **392**, 1–9.

Edwards, W.F. (1824) *De L'Influence Des Agens Physique Sur la Vie*. Crochard, Paris.

Edwards, M.J. & Martin, R.J. (1966) Mixing technique for the oxygen–hemoglobin equilibrium and Bohr effect. *Journal of Applied Physiology* **21**, 1898–1902.

Edwards, P.J., May, R.M. & Webb, N.R., eds. (1994) *Large-Scale Ecology and Conservation Biology*. Blackwell Scientific Publications, Oxford.

Ege, R. & Krogh, A. (1914) On the relation between the temperature and respiratory exchange in fishes. *Internationale Revue der Gesamen Hydrobiologie Hydrographie* **1**, 48–55.

Ehrlich, P.R., Breedlove, D.E., Brussard, P.F. & Sharp, M.A. (1972) Weather and the 'regulation' of subalpine populations. *Ecology* **53**, 243–247.

Ehrlich, P.R., Murphy, D.D., Singer, M.C. *et al.* (1980) Extinction, reduction, stability and increase: the responses of checkerspot butterfly (*Euphydryas*) populations to the California drought. *Oecologia* **46**, 101–105.

Eldredge, N. (1985) *Unfinished Synthesis: Biological Hierarchies and Modern Evolutionary Thought*. Oxford University Press, Oxford.

Elgar, M.A. & Harvey, P.H. (1987) Basal metabolic rates in mammals: allometry, phylogeny and ecology. *Functional Ecology* **1**, 25–36.

Elkins, N. (1995) *Weather and Bird Behaviour,* 2nd edn. Poyser, London.

Elliott, J.M. (1981) Some aspects of thermal stress of freshwater teleosts. In: *Stress and Fish* (ed. A.D. Pickering), pp. 209–245. Academic Press, London.

Elton, C.S. (1958) *The Ecology of Invasions by Animals and Plants.* Methuen, London.

Endler, J.A. (1986) *Natural Selection in the Wild.* Princeton University Press, Princeton, New Jersey.

Engel, D.W. & Vaughan, D.S. (1996) Biomarkers, natural variability, and risk assessment: Can they coexist? *Human Ecological Risk Assessment* **2**, 257–262.

Eppley, Z.A. (1994) A mathematical model of heat flux applied to developing endotherms. *Physiological Zoology* **67**, 829–854.

Eriksson, S.P. (1998) *Marine animal–sediment interactions: effects of hypoxia and manganese on the benthic crustacean,* Nephrops norvegicus *(L).* PhD Thesis, Göteborg University, Sweden.

Eriksson, S.P. & Baden, S.P. (1997) Behaviour and tolerance to hypoxia in juvenile Norway lobster (*Nephrops norvegicus*) of different ages. *Marine Biology* **128**, 49–54.

Etges, W.J. & Klassen, C.S. (1989) Influences of atmospheric ethanol on adult *Drosophila mojavensis*: altered metabolic rates and increases in fitness among populations. *Physiological Zoology* **62**, 170–193.

Evans, C.L. (1945) *Principles of Human Physiology,* 9th edn. J. & A. Churchill, London.

Facey, D.E. & Grossman, G.D. (1990) The metabolic cost of maintaining position for four North American stream fishes: effects of season and velocity. *Physiological Zoology* **63**, 757–776.

Faraci, F.M. (1991) Adaptations to hypoxia in birds: how to fly high. *Annual Review of Physiology* **53**, 59–70.

Farmanfarmaian, A. & Giese, A.G. (1963) Thermal tolerance and acclimation in the western purple sea urchin, *Strongylocentrotus purpuratus. Physiological Zoology* **36**, 237–243.

Fazekas, J.F., Alexander, F.A.D. & Himwich, H.E. (1941) Tolerance of the newborn to anoxia. *American Journal of Physiology* **134**, 281–287.

Feder, H.M. (1956) *Natural history studies on the starfish* Pisaster ochraceus *(Brandt 1835) in the Monterey Bay area.* PhD Thesis, Stanford University.

Feder, M.E. (1978) Environmental variability and thermal acclimation of metabolism in neotropical and temperate zone salamanders. *Physiological Zoology* **51**, 7–16.

Feder, M.E. (1981) Effect of body size, trophic state, time of day, and experimental stress on oxygen consumption of anuran larvae: an experimental assessment and evaluation of the literature. *Comparative Biochemistry and Physiology* **70A**, 497–508.

Feder, M.E. (1982a) Effect of developmental stage and body size on oxygen consumption of anuran larvae: a reappraisal. *Journal of Experimental Zoology* **220**, 33–42.

Feder, M.E. (1982b) Environmental variability and thermal acclimation of metabolism in tropical anurans. *Journal of Thermal Biology* **7**, 23–28.

Feder, M.E. (1987a) The analysis of physiological diversity: the prospects for pattern documentation and general questions in ecological physiology. In: *New Directions in Ecological Physiology* (eds M.E. Feder, A.F. Bennett, W.W. Burggren & R.B. Huey), pp. 38–75. Cambridge University Press, Cambridge.

Feder, M.E. (1987b) Effect of thermal acclimation on locomotor energetics and locomotor performance in a tropical salamander, *Bolitoglossa subpalmata. Physiological Zoology* **60**, 18–26.

Feder, M.E. (1992) A perspective on environmental physiology of the amphibians. In: *Environmental Physiology of the Amphibians* (eds M.E. Feder & W.W. Burggren), pp. 1–6, University of Chicago Press, Chicago.

Feder, M.E. (1996) Ecological and evolutionary physiology of stress proteins and the stress response: the *Drosophila melanogaster* model. In: *Animals and Temperature: Phenotypic and Evolutionary Adaptation* (eds I.A. Johnston & A.F. Bennett), pp. 79–102. Cambridge University Press, Cambridge.

Feder, M.E., Bennett, A.F., Burggren, W.W. & Huey, R.B., eds. (1987) *New Directions in Ecological Physiology.* Cambridge University Press, Cambridge.

Feder, M.E., Blair, N. & Figueras, H. (1997) Natural thermal stress and heat-shock protein expression in *Drosophila* larvae and pupae. *Functional Ecology* **11**, 90–100.

Feder, M.E. & Block, B. (1991) On the future of physiological ecology. *Functional Ecology* **5**, 136–144.

Feder, M.E. & Burggren, W.W., eds (1992) *Environmental Physiology of the Amphibians.* University of Chicago Press, Chicago.

Feder, M.E., Cartano, N.V., Milos, L., Krebs, R.A. & Lindquist, S.L. (1996) Effect of engineering Hsp70 copy number on Hsp70 expression and tolerance of ecologically relevant heat shock in larvae and pupae of *Drosophila melanogaster. Journal of Experimental Biology* **199**, 1837–1844.

Feder, M.E., Parsell, D.A. & Lindquist, S.L. (1995) The stress response and stress proteins. In: *Cell Biology of Trauma* (eds J.J. Lemasters & C. Oliver), pp. 177–191. CRC Press, Boca Raton, Florida.

Feibleman, J.K. (1955) Theory of integrative levels. *British Journal of the Philosophical Society* **5**, 59–66.

Feiler, R., Bjornson, R., Kirschfeld, K., Mismer, D., Rubin, G.M., Smith, D.P., Socolich, M. & Zuker, C.S. (1992) Ectopic expression of ultraviolet-rhodopsins in the blue photoreceptor cells of *Drosophila*: visual physiology and photochemistry of transgenic animals. *Journal of Neuroscience* **12**, 3862–3868.

Feist, D.D. & White, R.G. (1989) Terrestrial mammals in cold. *Advances in Computational Environmental Physiology* **4**, 327–360.

Felsenstein, J. (1985) Phylogenies and the comparative method. *American Naturalist* **125**, 1–15.

Ferguson, M.W.J. & Joanen, T. (1982) Temperature of egg incubation determines sex in *Alligator mississippiensis*. *Nature* **296**, 850–853.

Fields, P., Graham, J.B., Rosenblatt, R.H. & Somero, G.N. (1993) Effects of expected global change on marine faunas. *Trends in Ecology and Evolution* **8**, 30–37.

Fields, R., Lowe, S.S., Kaminski, C., Whitt, G.S. & Philipp, D.P. (1987) Critical and chronic thermal maxima of northern and Florida largemouth bass and their reciprocal F_1 and F_2 hybrids. *Transactions of the American Fisheries Society* **116**, 856–863.

Finch, C.E. (1990) *Longevity, Senescence and the Genome*. University of Chicago Press, Chicago.

Finn, J.B. (1937) Blood may tell. Popular adage confirmed in study of blood-cell size in salamander species. *Journal of Heredity* **28**, 373.

Florey, E. (1987) Fads and fallacies in contemporary physiology. In: *Advances in Physiological Research* (eds H. McLennan, J.R. Ledsome, C.H.S. McIntosh & D.R. Jones), pp. 91–110. Plenum Press, New York.

Flückiger, E. & Verzár, F. (1955) Lack of adaptation to low oxygen pressure in aged animals. *Journal of Gerontology* **10**, 306–311.

Flügel, H. (1972) Pressure: animals. In: *Marine Ecology, Vol 1: Environmental Factors* (Part 3) (ed. O. Kinne), pp. 1407–1450. Wiley-Interscience, London.

Forbes, A.T. & Hill, B.J. (1969) The physiological ability of a marine crab *Hymenosoma orbiculare* Desm. to live in a subtropical freshwater lake. *Transactions of the Royal Society of South Africa* **38**, 271–283.

Forbes, T.L. & Lopez, G.R. (1989) Determination of critical periods in ontogenic trajectories. *Functional Ecology* **3**, 625–632.

Forbes, V.E. & Forbes, T.L. (1994) *Ecotoxicology in Theory and Practice*. Chapman & Hall, London.

Forbes, V.E., Møller, V. & Depledge, M.H. (1995) Intrapopulation variability in sublethal response to heavy metal stress in sexual and asexual gastropod populations. *Functional Ecology* **9**, 477–484.

Forchhammer, M.C., Stenseth, N.C., Post, E. & Langvatn, R. (1998) Population dynamics of Norwegian red deer: density-dependence and climatic variation. *Proceedings of the Royal Society of London* **265B**, 341–350.

Fox, H.M. (1936) The activity and metabolism of poikilothermic animals in different latitudes. I. *Proceedings of the Zoological Society of London* **1936**, 945–955.

Fox, H.M. (1938) The activity and metabolism of poikilothermic animals in different latitudes. III. *Proceedings of the Zoological Society of London* **108A**, 501–505.

Fox, H.M. (1939) The activity and metabolism of poikilothermic animals in different latitudes. V. *Proceedings of the Zoological Society of London* **109A**, 141–156.

Fox, H.M. & Wingfield, C.A. (1937) The activity and metabolism of poikilothermic animals in different latitudes. II. *Proceedings of the Zoological Society of London* **1937** (Ser. A) 275–282.

Fritz, J.D. & Robertson, D. (1996) Gene targeting approaches to the autonomic nervous system. *Journal of Autonomic Nervous System* **61**, 1–5.

Fukamizu, A. & Murakami, K. (1997) Activated and inactivated renin-angiotensin system in transgenic animals: from genes to blood pressure. *Laboratory Animal Science* **47**, 127–131.

Funk, G.D. & Feldman, J.L. (1995) Generation of respiratory rhythm and pattern in mammals: insights from developmental studies. *Current Opinions in Neurobiology* **5**, 778–785.

Funk, G.D., Parkis, M.A., Selvaratnam, S.R. & Walsh, C. (1997) Developmental modulation of glutamatergic inspiratory drive to hypoglossal motoneurons. *Respiration Physiology* **110**, 125–137.

deFur, P.L., Mangum, C.P. & Reese, J.E. (1990) Respiratory responses of the blue crab *Callinectes sapidus* to long-term hypoxia. *Biological Bulletin* **178**, 46–54.

Futuyma, D.J. (1987) Interindividual comparisons: a discussion. In: *New Directions in Ecological Physiology* (eds M.E. Feder, A.F. Bennett, W.W. Burggren & R.B. Huey), pp. 240–247. Cambridge University Press, Cambridge.

Gage, J.D. (1986) The benthic fauna of the Rockall Trough: regional distribution and bathymetric zonation. *Proceedings of the Royal Society of Edinburgh* **88B**, 159–174.

Gaines, S.D. & Denny, M.W. (1993) The largest, smallest, highest, lowest, longest, and shortest: extremes in ecology. *Ecology* **74**, 1677–1692.

Gallivan, G.J. & Best, R.C. (1986) The influence of feeding and fasting on the metabolic rate and ventilation of the Amazonian manatee (*Trichechus inunguis*). *Physiological Zoology* **59**, 552–557.

Garland, T. Jr (1984) Physiological correlates of locomotory performance in a lizard: an allometric approach. *American Journal of Physiology* **247**, R808–R815.

Garland, T. Jr (1992) Rate tests for phenotypic evolution using phylogenetically independent contrasts. *American Naturalist* **140**, 509–519.

Garland, T. Jr & Adolph, S.C. (1991) Physiological differentiation of vertebrate populations. *Annual Review of Ecology and Systematics* **22**, 193–228.

Garland, T. Jr & Adolph, S.C. (1994) Why not to do two-species comparative studies: limitations on inferring adaptation. *Physiological Zoology* **67**, 797–828.

Garland, T. Jr & Carter, P.A. (1994) Evolutionary physiology. *Annual Review of Physiology* **56**, 579–621.

Garland, T. Jr & Else, P.L. (1987) Seasonal, sexual and individual variation in endurance and activity metabolism in lizards. *American Journal of Physiology* **252**, R439–R449.

Garland, T. Jr, Huey, R.B. & Bennett, A.F. (1991) Phylogeny and thermal physiology in lizards: a reanalysis. *Evolution* **45**, 1969–1975.

Garland, T.Jr, Martin, K.L.M. & Diaz-Uriate, R. (1997) Reconstructing ancestral trait values using squared-change parsimony: Plasma osmolarity at the origin of the amniotes. In: *Amniote Origins: Completing the Transition to Land* (eds S.S. Sumida & K.L.M. Martin), pp. 425–501. Academic Press, San Diego.

Garry, D.J., Ordway, G.A., Lorenz, J.N., Radford, N.B., Chin, E.R., Grange, R.W., Bassel-Duby, R. & Williams, R.S. (1998) Mice without myoglobin. *Nature* **395**, 905–908.

Garstang, W. (1922) The theory of recapitulation. A critical restatement of the biogenetic law. *Journal of Linnaean Society of London (Zoology)* **35**, 81–82.

Garton, D.W. (1984) Relationship between multiple locus heterozygosity and physiological energetics of growth in the estuarine gastropod, *Thais haemastoma*. *Physiological Zoology* **57**, 530–543.

Gaston, K.J. (1990) Patterns in the geographical ranges of species. *Biological Reviews* **65**, 105–129.

Gaston, K.J. (1994) *Rarity*. Chapman & Hall, London.

Gaston, K.J. (1996a) Species-range size distributions: patterns, mechanisms and implications. *Trends in Ecology and Evolution* **11**, 197–201.

Gaston, K.J. (1996b) The multiple forms of the interspecific abundance-distribution relationship. *Oikos* **75**, 211–220.

Gaston, K.J., ed. (1996c) *Biodiversity: A Biology of Numbers and Difference*. Blackwell Science, Oxford.

Gaston, K.J. (1998) Species-range size distributions: products of speciation, extinction and transformation. *Philosophical Transactions of the Royal Society of London* **353B**, 219–230.

Gaston, K.J. & Blackburn, T.M.A. (1999) critique for macroecology. *Oikos* **84**, 353–368.

Gaston, K.J., Blackburn, T.M. & Lawton, J.H. (1997) Interspecific abundance-range size relationships: an appraisal of mechanisms. *Journal of Animal Ecology* **66**, 579–601.

Gaston, K.J., Blackburn, T.M. & Spicer, J.I. (1998) Rapoport's rule: time for an epitaph? *Trends in Ecology and Evolution* **13**, 70–74.

Gaston, K.J. & McArdle, B.H. (1993) All else is not equal: temporal population variability and insect conservation. In: *Perspectives on Insect Conservation* (eds K.J. Gaston, T.R. New & M.J. Samways), pp. 171–184. Intercept, Andover.

Gaston, K.J. & Spicer, J.I. (1998a) *Biodiversity: an Introduction*. Blackwell Science, Oxford.

Gaston, K.J. & Spicer, J.I. (1998b) Do upper thermal tolerances differ in geographically-separated populations of the beachflea *Orchestia gammarellus* (Crustacea: Amphipoda)? *Journal of Experimental Marine Biology and Ecology* **229**, 265–276.

Gaston, K.J. & Williams, P.H. (1996) Spatial patterns in taxonomic diversity. In: *Biodiversity: a Biology of Numbers and Difference* (ed. K.J. Gaston), pp. 202–229. Blackwell Science, Oxford.

Gatten, R.E. Jr, Echternacht, A.C. & Wilson, M.A. (1988) Acclimatization versus acclimation of activity metabolism in a lizard. *Physiological Zoology* **61**, 322–329.

Gaunt, A.S., Hikida, R.S., Jehl, J.R. Jr & Fenbert, L. (1990) Rapid atrophy and hypertrophy of an avian flight muscle. *Auk* **107**, 649–659.

Geiger, R. (1965) *The Climate Near the Ground*. Harvard University Press, Cambridge, Massachussetts.

Geiser, F., Hiebert, S. & Kenagy, G.J. (1990) Torpor bout duration during the hibernation season of two sciurid rodents: interrelations with temperature and metabolism. *Physiological Zoology* **63**, 489–503.

Geiser, F. & Kenagy, G.J. (1988) Torpor duration in relation to temperature and metabolism in hibernating ground squirrels. *Physiological Zoology* **61**, 442–449.

Geiser, F. & Ruf, T. (1995) Hibernation versus daily torpor in mammals and birds: physiological variables and classification of torpor patterns. *Physiological Zoology* **68**, 935–966.

George, R.Y. (1981) Functional adaptations of deep sea organisms. In: *Functional Adaptations of Marine Organisms* (eds F.J. Vernberg & W.B. Vernberg), pp. 280–332. Academic Press, New York.

Gibbs, A.G., Louie, A.K. & Ayala, J.A. (1998) Effects of temperature on cuticular lipids and water balance in a desert *Drosophila*: is thermal acclimation beneficial? *Journal of Experimental Biology* **201**, 71–80.

Gibbs, A.G. & Mousseau, T.A. (1994) Thermal acclimation and genetic variation in cuticular lipids of the lesser migratory grasshopper (*Melanoplus sanguinipes*): effects of lipid composition on biophysical properties. *Physiological Zoology* **67**, 1523–1543.

Gilbert, N. (1980) Comparative dynamics of a single-host aphid. I. The evidence. *Journal of Animal Ecology* **49**, 351–369.

Gilles, R. & Delpire, E. (1997) Variations in salinity, osmolarity, and water availability. In: *Handbook of Physiology, Sect. 13: Comparative Physiology*, Vol 2 (ed. W.H. Dantzler), pp. 1523–1599. Oxford University Press, New York.

Gittleman, J.L. & Luh, H.-K. (1992) On comparing comparative methods. *Annual Review of Ecology and Systematics* **23**, 383–404.

Gnaiger, E., Lassnig, B., Kuznetsov, A., Rieger, G. & Margareiter, R. (1998) Mitochondrial oxygen affinity, respiratory flux control and excess capacity of cytochrome c oxidase. *Journal of Experimental Biology* **201**, 1129–1139.

Goodson, W.H. & Hunt, T.K. (1979) Wound healing and ageing. *Journal of Investigative Dermatology* **73**, 88–91.

Gordon, C.J. (1993) *Comparative Physiology of Temperature Regulation in Laboratory Rodents*. Cambridge University Press, Cambridge.

Gould, S.J. (1977) *Ontogeny and Phylogeny*. Harvard University Press, Cambridge, Massachussetts.

Gould, S.J. (1989) A developmental constraint in *Cerion*, with comments on the definition and interpretation of constraint in evolution. *Evolution* **43**, 516–539.

Gould, S.J. & Lewontin, R.C. (1979) The spandrels of San Marco and the panglossian paradigm: a critique of the adaptionist programme. *Proceedings of the Royal Society of London* **205B**, 581–598.

Grad, B. & Kral, V.A. (1957) The effects of senescence on resistance to stress. I. Responses of young and old mice to stress. *Journal of Gerontology* **12**, 172–181.

Grafen, A. (1989) The phylogenetic regression. *Philosophical Transactions of the Royal Society of London* **326B**, 119–157.

Graham, J.B. (1972) Low-temperature acclimation and the seasonal temperature sensitivity of some tropical marine fishes. *Physiological Zoology* **45**, 1–13.

Greengard, O., ed. (1974) *Biochemical Bases of the Development of Physiological Functions*. Karger, Farmington.

Greenwood, J.J.D. & Baillie, S.R. (1991) Effects of density-dependence and weather on population changes of English passerines using a non-experimental approach. *Ibis* **133**, 121–133.

Grenville, R.W. & Morgan, A.J. (1991) A comparison of lead, cadmium and zinc accumulation in terrestrial slugs maintained in microcosms: evidence for metal tolerance. *Environmental Pollution* **74**, 115–128.

Griffith, B., Scott, J.M., Carpenter, J.W. & Reed, C. (1989) Translocation as a species conservation tool: status and strategy. *Science* **245**, 477–480.

Grosholz, E.D. & Ruiz, G.M. (1996) Predicting the impact of introduced marine species: lessons from the multiple invasions of the European green crab *Carcinus maenas*. *Biological Conservation* **78**, 59–66.

Grueber, W.B. & Bradley, T.J. (1994) The evolution of increased salinity tolerance in larvae of *Aedes* mosquitoes: a phylogenetic analysis. *Physiological Zoology* **67**, 566–579.

Gunter, G. (1957) Temperature. In: *Treatise on Marine Ecology and Paleoecology*, Vol 1: *Ecology* (ed. J.W. Hedgepeth). *Memoirs of the Geological Society of America* **67**, 159–184.

Gutzke, W.H.N. & Crews, D. (1988) Embryonic temperature determines adult sexuality in a reptile. *Nature* **332**, 832–834.

Hadley, N.F. & Massion, D.D. (1985) Oxygen consumption, water loss and cuticular lipids of high and low elevation populations of the grasshopper *Aeropedellus clavatus* (Orthoptera: Acrididae). *Comparative Biochemistry and Physiology* **80A**, 307–311.

Haeckel, E. (1866) *Generelle Morphologie der Organismen: Allgemeine Grundzüge der Organischen Formen-Wissenschaft, Mechanisch Begründet durch die von Charles Darwin Reformite Descendenz-Theorie* (2 volumes). George Reimer, Berlin.

Haeckel, E. (1876) *The history of creation: or, the development of the earth and its inhabitants by the action of natural causes: a popular exposition of the doctrine of evolution in general and that of Darwin, Goethe, and Lamarck in particular* (2 volumes). Appleton, New York.

Hagerman, L. (1986) Haemocyanin concentration in the shrimp *Crangon crangon* (L.) after exposure to moderate hypoxia. *Comparative Biochemistry and Physiology* **85A**, 721–724.

Hagerman, L. & Baden, S.P. (1988) *Nephrops norvegicus*: field study on the effects of oxygen deficiency on haemocyanin concentration. *Journal of Experimental Marine Biology and Ecology* **116**, 135–142.

Hagerman, L. & Uglow, R.F. (1985) Effects of hypoxia on the respiratory and circulatory regulation of *Nephrops norvegicus*. *Marine Biology* **87**, 273–278.

Haila, Y. & Hanski, I.K. (1993) Birds breeding on small British islands and extinction risks. *American Naturalist* **142**, 1025–1029.

Haim, A., Heth, G., Avnon, Z. & Nevo, E. (1984) Adaptive physiological variation in non-shivering thermogenesis and its significance in speciation. *Journal of Comparative Physiology* **154B**, 145–147.

Haim, A., Heth, G. & Nevo, E. (1985) Adaptive thermoregulatory pattern in speciating mole rats. *Acta Zoologica Fennica* **170**, 137–140.

Hainsworth, F.R. (1981) *Animal Physiology: Adaptation in Function*. Addison-Wesley-Longman, New York.

Hall, B.K. (1992) *Evolutionary Developmental Biology*. Chapman & Hall, London.

Hall, F.G., Dill, D.B. & Guzman-Barron, E.S. (1936) Comparative physiology in high altitudes. *Journal of Cellular and Comparative Physiology* **8**, 301–313.

Haller, J. & Wittenberger, C. (1988) Biochemical energetics of hierarchy formation in *Betta splendens*. *Physiological Behavior* **43**, 447–450.

Hammond, K.A. & Diamond, J. (1992) An experimental test for a ceiling on sustained metabolic rate in lactating mice. *Physiology and Zoology* **65**, 952–977.

Hammond, K.A. & Diamond, J. (1994) Limits to dietary nutrient intake and intestinal nutrient uptake in lactating mice. *Physiological Zoology* **67**, 282–303.

Hammond, K.A. & Diamond, J. (1997) Maximal sustained energy budgets in humans and animals. *Nature* **386**, 457–462.

Hammond, K.A., Konarzewski, M., Torres, R.M. & Diamond, J. (1994) Metabolic ceilings under a combination of peak energy demands. *Physiological Zoology* **67**, 1479–1506.

Hammond, K.A., Lloyd, K.C.K. & Diamond, J. (1996) Is mammary output capacity limiting to lactational performance in mice? *Journal of Experimental Biology* **199**, 337–349.

Hansen, S. & Lavigne, D.M. (1997) Ontogeny of thermal limits in the harbor seal (*Phoca vitulina*). *Physiological Zoology* **70**, 85–92.

Hanski, I. (1982) Dynamics of regional distribution: the core and satellite species hypothesis. *Oikos* **38**, 210–221.

Hanski, I. & Gilpin, M.E., eds. (1997) *Metapopulation Biology: Ecology, Genetics, and Evolution*. Academic Press, San Diego.

Hardison, R. (1998) Hemoglobins from bacteria to man: evolution of different patterns of gene expression. *Journal of Experimental Biology* **201**, 1099–1117.

Harris, R.R. & Aladin, N.V. (1997) The ecophysiology of osmoregulation in Crustacea. In: *Ionic Regulation in Animals: A Tribute to Professor W.T.W. Potts* (ed. D.H. Evans), pp. 1–25. Springer-Verlag, Berlin.

Harrison, S., Murphy, D.D. & Ehrlich, P.R. (1988) Distribution of the bay checkerspot butterfly, *Euphydryas editha bayensis*: evidence for a metapopulation model. *American Naturalist* **132**, 360–382.

Hart, J.S. (1964) Geography and season: mammals and birds. In: *Handbook of Physiology, Sect. 4: Adaptations to the Environment* (eds D.B. Dill, E.F. Adolph & C.G. Wilber), pp. 283–294. American Physiological Society, Washington, DC.

Harvey, W. (1628) *Exercitatio Anatomica de Motu Cordis et Sanguinis in Animalibu*, Frankfurt. (An anatomical disputation concerning the movement of the heart and blood in living creatures.) Translated by G. Whitteridge (1976). Blackwell, Oxford.

Harvey, P.H. (1996) Phylogenies for ecologists. *Journal of Animal Ecology* **65**, 255–263.

Harvey, P.H. & Pagel, M.D. (1991) *The Comparative Method in Evolutionary Biology*. Oxford University Press, Oxford.

Haschemi, H. (1992) Studies of the cold tolerance of different strains of the German cockroach (*Blattella germanica* (L.), Blattodea, Blattidae) under laboratory conditions. *Zeitschrift für Angewandte Zoologie* **79**, 335–348.

Hastings, D. & Burggren, W.W. (1995) Developmental changes in oxygen consumption regulation in larvae of the South African clawed toad *Xenopus laevis*. *Journal of Experimental Biology* **198**, 2465–2475.

Hatch, T. (1962) Changing objectives in occupational health. *Industrial Hygiene Journal* **62**, 1–7.

Hawkins, A.J.S., Bayne, B.L., Day, A.J., Rus, J. & Worrall, C.M. (1989) Genotype dependent interrelations between energy metabolism, protein metabolism and fitness. In: *Reproduction, Genetics and Distribution of Marine Organisms* (eds J. Ryland & P.A. Tyler), pp. 283–292. Olsen & Olsen, Fredensberg.

Hawkins, B.A. & Holyoak, M. (1998) Transcontinental crashes of insect populations? *American Naturalist* **152**, 480–484.

Hawksworth, D.L. & Kalin-Arroyo, M.T. (1995) Magnitude and distribution of biodiversity. In: *Global Biodiversity Assessment* (ed. V.H. Heywood), pp. 107–191. Cambridge University Press, Cambridge.

Hayes, J.P. (1989) Field and maximal metabolic rates of deer mice (*Peromyscus maniculatus*) at low and high altitudes. *Physiological Zoology* **62**, 732–744.

Hayes, J.P., Bible, C.A. & Boone, J.D. (1998) Repeatability of mammalian physiology: evaporative water loss and oxygen consumption of *Dipodomys merriami*. *Journal of Mammalogy* **79**, 475–485.

Hayes, J.P., Garland, T. Jr & Dohm, M.R. (1992) Individual variation in metabolism and reproduction of *Mus*: are energetics and life history linked? *Functional Ecology* **6**, 5–15.

Hayes, J.P. & Shonkwiler, J.S. (1996) Altitudinal effects on water fluxes of deer mice: a physiological application of structural equation modeling with latent variables. *Physiological Zoology* **69**, 509–531.

Hayssen, V. & Lacy, R.C. (1985) Basal metabolic rates in mammals: taxonomic differences in the allometry of BMR and body mass. *Comparative Biochemistry and Physiology* **81A**, 741–754.

Hazel, J.R. (1995) Thermal adaptation in biological membranes: is homeoviscous adaptation the explanation? *Annual Review of Physiology* **57**, 19–42.

Heatwole, H., Mercado, N. & Ortiz, E. (1965) Comparison of critical thermal maxima of two species of Puerto Rican frogs of the genus *Eleutherodactylus*. *Physiological Zoology* **38**, 1–8.

Heatwole, H. & Muir, R. (1989) Seasonal and daily activity of ants in the pre-Saharan steppe of Tunisia. *Journal of Arid Environment* **16**, 49–67.

Hengeveld, R. (1989) *Dynamics of Biological Invasions*. Chapman & Hall, London.

Henning, S.J. (1985) Ontogeny of enzymes in the small intestine. *Annual Review of Physiology* **47**, 231–245.

Hepper, P.G. (1987) The amniotic fluid: an important priming role in kin recognition. *Animal Behavior* **35**, 1343–1346.

van Herrewege, J. & David, J.R. (1997) Starvation and desiccation tolerances in *Drosophila*: comparison of species from different climatic origins. *Écoscience* **4**, 151–157.

Hesse, R., Allee, W.C. & Schmidt, K.P. (1951) *Ecological Animal Geography*, 2nd edn. John Wiley, New York.

Heusner, A.A. (1982) Energy metabolism and body size. 1. Is the 0.75 mass exponent of Kleiber's equation a statistical artefact? *Respiration Physiology* **48**, 1–12.

Hewatt, W.G. (1937) Ecological studies on selected marine intertidal communities of Monterey Bay, California. *American Midland Naturalist* **18**, 161–206.

Heywood, V.H., ed. (1995) *Global Biodiversity Assessment*. Cambridge University Press, Cambridge.

Hilbish, T.J. (1985) Demographic and temporary structure of an allele frequency cline in the mussel *Mytilus edulis*. *Marine Biology* **86**, 163–172.

Hilbish, T.J., Bayne, B.L. & Day, A. (1994) Genetics of physiological differentiation within the marine mussel genus *Mytilus*. *Evolution* **48**, 267–286.

Hilbish, T.J., Denton, L.E. & Koehn, R.K. (1982) Effect of an allozyme polymorphism on regulation of cell volume. *Nature* **298**, 688–689.

Hilbish, T.J. & Koehn, R.K. (1985a) The physiological basis of natural selection at the *Lap* locus. *Evolution* **39**, 1302–1307.

Hilbish, T.J. & Koehn, R.K. (1985b) Dominance in physiological phenotypes and fitness at an enzyme locus. *Science* **229**, 52–54.

Hill, R.W. (1976) *Comparative Physiology of Animals: An Environmental Approach*. Harper & Row, New York.

Himwich, H.E., Bernstein, A.D., Herrlich, H., Chesler, A. & Fasekan, J.F. (1942) Mechanisms for maintenance of life in newborn during anoxia. *American Journal of Physiology* **135**, 387–391.

Hlobowsky, J.I. & Wissing, T.E. (1985) Seasonal changes in critical thermal maxima of fantail (*Etheostoma flabellare*), greenside (*Etheostoma blennioides*) and rainbow (*Etheostoma caeruleum*) darters. *Canadian Journal of Zoology* **63**, 1629–1633.

Hoar, W.S. (1983) *General and Comparative Physiology*. Prentice-Hall, London.

Hoar, W.S. (1988) The physiology of smolting salmonids. In: *Fish Physiology,* Vol XIB (eds W.S. Hoar & D.J. Randall), pp. 275–344. Academic Press, New York.

Hochachka, P.W. & Guppy, M. (1987) *Metabolic Arrest and the Control of Biological Time*. Harvard University Press, Cambridge, Massachusetts.

Hochachka, P.W. & Somero, G.N. (1973) *Strategies of Biochemical Adaptation*. Saunders, Pennsylvania.

Hodkinson, I.K. (1997) Progressive restriction of host plant exploitation along a climatic gradient: the willow psyllid *Cacopsylla groenlandica* in Greenland. *Ecological Entomology* **22**, 47–54.

Hoffmann, A.A. (1995) The cost of acclimation. *Trends in Ecology and Evolution* **10**, 1–2.

Hoffmann, A.A. & Blows, M.W. (1993) Evolutionary genetics and climate change: will animals adapt to global warming? In: *Biotic Interactions and Global Change* (eds P.M. Kareiva, R.B. Huey & J.G. Kingsolver), pp. 165–178. Sinauer Associates, Sunderland, Massachusetts.

Hoffmann, A.A. & Blows, M.W. (1994) Species borders: ecological and evolutionary perspectives. *Trends in Ecology and Evolution* **9**, 223–227.

Hoffmannn, A.A. & Parsons, P.A. (1991) *Evolutionary Genetics and Environmental Stress*. Oxford University Press, Oxford.

Hoffmann, A.A. & Parsons, P.A. (1997) *Extreme Environmental Change and Evolution*. Cambridge University Press, Cambridge.

Hokkanen, J.E.I. & Demont, M.E. (1997) Complex dynamics in the heart of the lobster *Homarus americanus*. *Canadian Journal of Zoology* **75**, 746–754.

Holeton, G.F. (1974) Metabolic cold adaptation of polar fish: fact or artifact? *Physiological Zoology* **47**, 137–152.

Holland, L.Z., McFall, N.M. & Somero, G.N. (1997) Evolution of lactate dehydrogenase-A homologs of barracuda fishes (genus *Sphyraena*) from different thermal environments: differences in kinetic properties and thermal stability are due to amino acid substitutions outside the active site. *Biochemistry* **36**, 3207–3215.

Holland, W.E., Smith, M.H., Gibbons, J.W. & Brown, D.H. (1974) Thermal tolerances of fish from a reservoir receiving heated effluent from a nuclear reactor. *Physiological Zoology* **47**, 110–118.

Holliday, F.G.T. & Blaxter, J.H.S. (1960) The effects of salinity on the developing eggs and larvae of the herring. *Journal of the Marine Biological Association of the UK* **39**, 591–603.

Hopkins, H.S. (1946) The influence of season, concentration of sea water and environmental temperature upon the oxygen consumption of tissues in *Venus mercenaria*. *Journal of Experimental Zoology* **102**, 143–158.

Hoppeler, H. & Weibel, E.R. (1998) Limits for oxygen and substrate transport in mammals. *Journal of Experimental Biology* **201**, 1051–1064.

Horwood, J.W. & Millner, R.S. (1998) Cold induced abnormal catches of sole. *Journal of the Marine Biological Association of the UK* **78**, 345–347.

Hou, P.-C.L. & Burggren, W.W. (1989) Interaction of allometry and development in the mouse *Mus musculus*: heart rate and hematology. *Respiration Physiology* **78**, 265–280.

Houlihan, D.F. & Allan, D. (1982) Oxygen consumption of some Antarctic and British gastropods: an evaluation of cold adaptation. *Comparative Biochemistry and Physiology* **73A**, 383–387.

Houlihan, D.F., Waring, C.P., Mathers, E. & Gray, C. (1990) Protein synthesis and oxygen consumption of the shore crab *Carcinus maenas* after a meal. *Physiological Zoology* **63**, 735–756.

Hourdry, J., L'Hermite, A. & Ferrand, R. (1996) Changes in the digestive tract and feeding behavior of anuran amphibians during metamorphosis. *Physiological Zoology* **69**, 219–251.

Howe, R.W. (1958) A theoretical evaluation of the potential range and importance of *Trogoderma granarium* Everts in North America (Col. Dermestidae). *Proceedings of the 10th International Congress on Entomology* **4**, 23–28.

Howe, R.W. & Lindgren, D.L. (1957) How much can the khapra beetles spread in the USA? *Journal of Economic Entomology* **50**, 374–375.

Huey, R.B. (1991) Physiological consequences of habitat selection. *American Naturalist* **137**, S91–S115.

Huey, R.B. & Bennett, A.F. (1990) Physiological adjustments to fluctuating thermal environments: an ecological and evolutionary perspective. In: *Stress Proteins in Biology and Medicine* (eds R.I. Morimoto, A. Tissieres & C. Georgopoulus), pp. 37–59. Cold Spring Harbor Laboratory Press, Cold Spring Harbor, New York.

Huey, R.B. & Berrigan, D. (1996) Testing evolutionary hypotheses of acclimation. In: *Animals and Temperature: Phenotypic and Evolutionary Adaptation* (eds I.A. Johnston & A.F. Bennett), pp. 205–237. Cambridge University Press, Cambridge.

Huey, R.B. & Dunham, A.E. (1987) Repeatability of locomotor performance in natural populations of the lizard *Sceloporus merriami*. *Evolution* **41**, 1116–1120.

Huey, R.B. & Kingsolver, J.G. (1989) Evolution of thermal sensitivity of ectotherm performance. *Trends in Ecology and Evolution* **4**, 131–135.

Huey, R.B. & Kingsolver, J.G. (1993) Evolution of resistance to high temperatures in ectotherms. *American Naturalist* **142** (Suppl.), S21–S46.

Huey, R.B. & Stevenson, R.D. (1979) Integrating thermal physiology and ecology of ectotherms: a discussion of approaches. *American Zoologist* **19**, 357–366.

Hughes, L., Cawsey, E.M. & Westoby, M. (1996) Geographic and climatic range sizes of Australian eucalypts and a test of Rapoport's rule. *Global Ecology and Biogeography Letters* **5**, 128–142.

Hughes, R.N. & Roberts, D.J. (1980) Growth and reproductive rates in *Littorina neritoides* (L.) in North Wales. *Journal of the Marine Biological Association of the UK* **60**, 591–599.

Hunter, M.L. Jr (1996) *Fundamentals of Conservation Biology*. Blackwell Science, Oxford.

Huntsman, A.G. (1925) Limiting factors for marine animals. 2. Resistance of larval lobsters to extreme temperatures. *Contributions to Canadian Biology* **2**, 91–93.

Huntsman, A.G. & Sparkes, M.I. (1925) Limiting factors for marine animals. 3. Relative resistance to high temperatures. *Contributions to Canadian Biology* **2**, 97–113.

Hutchins, L.W. (1947) The bases for temperature zonation in geographical distribution. *Ecological Monographs* **17**, 325–335.

Huxley, J. (1942) *Evolution: The Modern Synthesis*. George Allen & Unwin, London.

Hyman, L.H. (1919) Physiological studies on *Planaria*. I. Oxygen consumption in relation to feeding and starvation. *American Journal of Physiology* **49**, 377–402.

Hyman, L.H. (1955) *The Invertebrates: Echinodermata*. McGraw-Hill, New York.

Hynes, H.B.N. (1954) The ecology of *Gammarus duebeni* Lilljeborg and its occurrence in freshwater in Western Britain. *Journal of Animal Ecology* **23**, 38–84.

Ikeda, T. (1988) Metabolism and chemical composition of crustaceans from the Antarctic mesopelagic zone. *Deep-Sea Research* **35A**, 1991–2002.

Jackson, S. & Diamond, J. (1995) Ontogenetic development of gut function, growth, and metabolism in a wild bird, the Red Jungle Fowl. *American Journal of Physiology* **269**, R1163–R1173.

Jacobs, L.F. (1996) The economy of winter: phenotypic plasticity in behavior and brain structure. *Biological Bulletin* **191**, 92–100.

Jacobson, M. (1991) *Developmental Neurobiology*, 3rd edn. Plenum, New York.

Jacobson, E.R. & Whitfield, W.G. (1970) The effect of acclimation on physiological responses to temperature in the snakes *Thamnophis proximus* and *Natrix rhombifera*. *Comparative Biochemistry and Physiology* **35**, 439–449.

Janes, D.N. & Chappell, M.A. (1995) The effect of ration size and body size on specific dynamic of action in Adelie penguin chicks, *Pygoscelis adeliae*. *Physiological Zoology* **68**, 1029–1044.

Jansen, A.H. & Chernick, V. (1983) Development of respiratory control. *Physiological Reviews* **63**, 437–483.

Jeffree, C.E. & Jeffree, E.P. (1996) Redistribution of the potential geographic ranges of mistletoe and Colorado beetle in Europe in response to the temperature component of climate change. *Functional Ecology* **10**, 562–577.

Jeffree, E.P. & Jeffree, C.E. (1994) Temperature and the biogeographical distribution of species. *Functional Ecology* **8**, 640–650.

Jessen, T.H., Weber, R.E., Fermi, G., Tame, J. & Braunitzer, G. (1991) Adaptation of bird hemoglobins to high altitudes: demonstration of molecular mechanism by protein engineering. *Proceedings of the National Academy of Sciences of the USA* **88**, 6519–6522.

Jobling, M. (1981) The influences of feeding on the metabolic rate of fishes: a short review. *Journal of Fish Biology* **18**, 385–400.

Jobling, M. & Davies, P.S. (1980) Effects of feeding on the metabolic rate and the specific

dynamic action in plaice, *Pleuronectes platessa* L. *Journal of Fish Biology* **16**, 629–638.

Johannesson, K., Kautsky, N. & Tedengren, M. (1990) Genotypic and phenotypic differences between Baltic and North Sea populations of the *Mytilus edulis* complex evaluated through reciprocal transplantations. II. Genetic variation. *Marine Ecology—Progress Series* **59**, 211–219.

Johansen, K. (1970) Air breathing in fishes. In: *Fish Physiology,* Vol IV (eds W.S. Hoar & D.J. Randall), pp. 361–413. Academic Press, New York.

John-Alder, H.B., McMann, S., Katz, L.S., Gross, A. & Barton, D.S. (1996) Social modulation of exercise endurance in a lizard (*Sceloporus undulatus*). *Physiological Zoology* **69**, 547–567.

Johnson, G. (1979) Genetic polymorphism among enzyme loci. In: *Physiological Genetics* (ed. J.G. Scandalios), pp. 239–280. Academic Press, New York.

Johnson, M.S. (1971) Adaptive lactate dehydrogenase variation in the crested blenny *Anoplarchus*. *Heredity* **27**, 205–226.

Johnson, K.S., Childress, J.J. & Beehler, C.L. (1988) Short-term temperature variability in the Rose Garden hydrothermal vent field: an unstable deep-sea environment. *Deep-Sea Research* **35**, 1711–1721.

Johnston, I.A. & Bennett, A.F., eds. (1996) *Animals and Temperature Phenotypic and Evolutionary Adaptation*. Cambridge University Press, Cambridge.

Johnston, I.A., Calvo, J., Guderley, H., Fernandez, D. & Palmer, L. (1998) Latitudinal variation in the abundance and oxidative capacities of muscle mitochondria in perciform fishes. *Journal of Experimental Biology* **201**, 1–12.

Johnston, I.A., Clarke, A. & Ward, P. (1991) Temperature and metabolic rate in sedentary fish from the Antarctic, North Sea and Indo-West Pacific Ocean. *Marine Biology* **109**, 191–195.

Johnston, I.A., Cole, N.J., Vieira, V.L.A. & Davidson, I. (1997) Temperature and developmental plasticity of muscle phenotype in herring larvae. *Journal of Experimental Biology* **200**, 849–868.

Johnston, I.A., Guderley, H., Franklin, C.E., Crockford, T. & Kamunde, C. (1994) Are mitochondria subject to evolutionary temperature adaptation? *Journal of Experimental Biology* **195**, 293–306.

Johnston, I.A., Vieira, V.L.A. & Hill, J. (1996) Temperature and ontogeny in ectotherms: muscle phenotype in fish. In: *Animals and Temperature: Phenotypic and Evolutionary Adaptations* (eds I.A. Johnston & A.F. Bennett), pp. 153–181. Cambridge University Press, Cambridge.

Jones, G. (1987) Selection against large size in

the Sand Martin *Riparia riparia* during a dramatic population crash. *Ibis* **129**, 274–280.

Jones, J.H. & Lindstedt, S.L. (1993) Limits to maximal performance. *Annual Review of Physiology* **55**, 547–569.

Jones, M.E., Grigg, G.C. & Beard, L.A. (1997) Body temperatures and activity patterns of Tasmanian devils (*Sarcophilus harrisii*) and eastern quolls (*Dasyurus viverrinus*) through a subalpine winter. *Physiological Zoology* **70**, 53–60.

Jonker, F.H., Van Oord, H.A., Van Geijn, H.P., Van Der Weijden, G.C. & Taverne, M.A.M. (1994) Feasibility of continuous recording of fetal heart rate in the near term bovine fetus by means of transabdominal Doppler. *Veterinary Quarterly* **16**, 165–168.

Kamada, T. (1933) The vapour pressure of the blood of edible snail. *Journal of Experimental Biology* **10**, 75–78.

Kareiva, P.M., Kingsolver, J.G. & Huey, R.B., eds. (1993) *Biotic Interactions and Global Change*. Sinaeur Associates, Sunderland, Massachusetts.

Kato, Y. (1959) The induction of phosphatase in various organs of the chick embryo. *Developmental Biology* **1**, 447–510.

Kavaliers, M. (1990) Responsiveness of deer mice to a predator, the short-tailed weasel: population differences and neuromodulatory mechanisms. *Physiological Zoology* **63**, 388–407.

Kelley, B.J. & Burbanck, W.D. (1976) Responses of embryonic *Cyathura polita* (Stimpson) (Isopoda: Anthuridae) to varying salinities. *Chesapeake Science* **17**, 159–167.

Kendeigh, S.C. (1939) The relation of metabolism to the development of temperature regulation in birds. *Journal of Experimental Zoology* **82**, 419–438.

Kennett, R. & Christian, K. (1994) Metabolic depression in estivating long-neck turtles (*Chelodina rugosa*). *Physiological Zoology* **67**, 1087–1102.

Kennington, G.S. (1957) Influence of altitude and temperature upon rate of oxygen consumption of *Tribolium confusum* Duval and *Camponotus pennsylvanicus modoc* Wheeler. *Physiological Zoology* **30**, 305–314.

Kent, J., Prosser, C.L. & Graham, G. (1992) Alterations in liver composition of channel catfish (*Ictalurus punctatus*) during seasonal acclimatization. *Physiological Zoology* **65**, 867–884.

Kerkut, G.A. & Meech, R.W. (1967) The effects of ions on the membrane potential of snail neurones. *Comparative Biochemistry and Physiology* **20**, 411–429.

Kersten, M. & Piersma, T. (1987) High levels of energy expenditure in shorebirds: metabolic adaptations to an energetically expensive way of life. *Ardea* **75**, 175–187.

Kingsolver, J.G. (1985) Butterfly thermoregulation: organismic mechanisms and population consequences. *Journal of Research into Lepidoptera* **24**, 1–20.

Kingsolver, J.G. (1989) Weather and population dynamics of insects: integrating physiological and population ecology. *Physiological Zoology* **62**, 314–334.

Kingsolver, J.G. (1995a) Viability selection on seasonally polyphenic traits: wing melanin pattern in Western White butterflies. *Evolution* **49**, 932–941.

Kingsolver, J.G. (1995b) Fitness consequences of seasonal polyphenism in Western White butterflies. *Evolution* **49**, 942–954.

Kingsolver, J.G. (1996) Experimental manipulation of wing pigment pattern and survival in Western White butterflies. *American Naturalist* **147**, 296–306.

Kingsolver, J.G. & Huey, R.B. (1998) Evolutionary analyses of morphological and physiological plasticity in thermally variable environments. *American Zoologist* **38**, 545–560.

Kingsolver, J.G., Huey, R.B. & Kareiva, P.M. (1993) An agenda for population and community research on global change. In: *Biotic Interactions and Global Change* (eds P.M. Kareiva, R.B. Huey & J.G. Kingsolver), pp. 480–486. Sinauer Associates, Sunderland, Massachusetts.

Kingsolver, J.G. & Watt, W.B. (1983) Thermoregulatory strategies in *Colias* butterflies: thermal stress and the limits to adaptation in temporally varying environments. *American Naturalist* **121**, 32–55.

Kingsolver, J.G. & Wiernasz, D.C. (1991) Seasonal polyphenism in wing-melanin pattern and thermoregulatory adaptation in *Pieris* butterflies. *American Naturalist* **137**, 816–830.

Kinne, O. (1962) Irreversible non-genetic adaptation. *Comparative Biochemistry and Physiology* **5**, 265–282.

Kinne, O. (1970) Invertebrates. Temperature effects. In: *Marine Ecology*, Vol 1: *Environmental Factors*, Part 1 (ed. O. Kinne), pp. 407–514. Wiley-Interscience, Chichester.

Kirby, R.R., Bayne, B.L. & Berry, R.J. (1994) Physiological variation on the dog-whelk *Nucella lapillus* L. either side of a cline in allozyme and karyotype frequencies. *Biological Journal of the Linnaean Society* **53**, 277–290.

Kirkpatrick, M. & Barton, N.H. (1997) Evolution of a species' range. *American Naturalist* **150**, 1–23.

Kirkpatrick, M. & Lofsvold, D. (1992) Measuring selection and constraint in the evolution of growth. *Evolution* **46**, 954–971.

Kirkwood, T.B.L. (1981) Repair and its

evolution: survival versus reproduction. In: *Physiological Ecology: An Evolutionary Approach to Resource Use* (eds C.R. Townsend & P. Calow), pp. 165–189. Blackwell Scientific Publications, Oxford.

Kirschner, L.B. (1967) Comparative physiology: invertebrate excretory organs. *Annual Review of Physiology* **29**, 169–196.

Kleckner, N.W. & Sidell, B.D. (1985) Comparison of maximal activities of enzymes from tissues of thermally acclimated and naturally acclimatized chain pickerel (*Esox niger*). *Physiological Zoology* **58**, 18–28.

Kleiber, M. (1932) Body size and metabolism. *Hilgardia* **6**, 315–353.

Kleiber, M. (1947) Body size and metabolic rate. *Physiological Review* **27**, 511–541.

Kleiber, M. (1961) *The Fire of Life: An Introduction to Animal Energetics*. John Wiley, New York.

Knapp, J.R. & Kopchick, J. (1994) The use of transgenic mice in nutrition research. *Journal of Nutrition* **124**, 461–468.

Knight-Jones, E.W. & Morgan, E. (1966) Responses of marine animals to changes in hydrostatic pressure. *Oceanography and Marine Biology Annual Review* **4**, 267–300.

Koehn, R.K. (1983) Biochemical genetics and adaptation in molluscs. In: *The Mollusca,* Vol 2 (ed. P.W. Hochachka), pp. 305–330. Academic Press, New York.

Koehn, R.K. (1987) The importance of genetics to physiological ecology. In: *New Directions in Ecological Physiology* (eds M.E. Feder, A.F. Bennett, W.W. Burggren & R.B. Huey), pp. 170–185. Cambridge University Press, Cambridge.

Konarzewski, M. & Diamond, J. (1994) Peak sustained metabolic rate and its individual variation in cold-stressed mice. *Physiological Zoology* **67**, 1186–1212.

Konarzewski, M. & Diamond, J. (1995) Evolution of basal metabolic rate and organ masses in laboratory mice. *Evolution* **49**, 1239–1248.

Kopp, J.B. & Klotman, P.E. (1995) Transgenic animal models of renal development and pathogenesis. *American Journal of Physiology* **269**, F601–F620.

Korsmeyer, K.E., Lai, N.C., Shadwick, R.E. & Graham, J.B. (1997) Heart rate and stroke volume contributions to cardiac output in swimming yellowfish tuna; response to exercise and temperature. *Journal of Experimental Biology* **200**, 1975–1986.

Koteja, P. (1996) Limits to the energy budget in a rodent, *Peromyscus maniculatus*: does gut capacity set the limit? *Physiological Zoology* **69**, 994–1020.

Kozlowski, J. & Weiner, J. (1997) Interspecific allometries are by-products of body size

optimization. *American Naturalist* **149**, 352–380.

Krebs, C.J. (1972) *Ecology: the Experimental Analysis of Distribution and Abundance*. Harper & Row, New York.

Krebs, H.A. (1975) The August Krogh principle: 'For many problems there is an animal on which it can be most conveniently studied'. *Journal of Experimental Zoology* **194**, 309–344.

Krebs, R.A. & Feder, M.E. (1997a) Natural variation in the expression of the heat-shock protein Hsp70 in a population of *Drosophila melanogaster* and its correlation with tolerance of ecologically relevant thermal stress. *Evolution* **51**, 173–179.

Krebs, R.A. & Feder, M.E. (1997b) Tissue-specific variation in Hsp70 expression and thermal damage in *Drosophila melanogaster* larvae. *Journal of Experimental Biology* **20**, 2007–2015.

Krebs, R.A. & Feder, M.E. (1997c) Deleterious consequences of Hsp70 overexpression in *Drosophila melanogaster* larvae. *Cell Stress and Chaperones* **2**, 60–71.

Krogh, A. (1916) *Respiratory Exchange of Animals and Man*. Longmans, London.

Krogh, A. (1939) *Osmotic Regulation in Aquatic Animals*. Cambridge University Press, Cambridge.

Kukal, O., Ayres, M.P. & Scriber, J.M. (1991) Cold tolerance of the pupae in relation to the distribution of swallowtail butterflies. *Canadian Journal of Zoology* **69**, 3028–3037.

Kunin, W.E. & Gaston, K.J. (1993) The biology of rarity: patterns, causes, and consequences. *Trends in Ecology and Evolution* **8**, 298–301.

Kunin, W.E. & Gaston, K.J., eds. (1997) *The Biology of Rarity: Causes and Consequences of Rare–Common Differences*. Chapman & Hall, London.

Laming, P.R., Funston, C.W., Roberts, D. & Armstrong, M.J. (1982) Behavioral, physiological and morphological adaptations of the shanny (*Blennius pholis*) to the intertidal habitat. *Journal of the Marine Biological Association of the UK* **62**, 329–338.

Landois, L. (1885) *A Text Book of Human Physiology* (2 volumes). Charles Griffin, London.

Lange, B.W., Langley, C.H. & Stephan, W. (1990) Molecular evolution of *Drosophila* metallothionein genes. *Genetics* **126**, 921–932.

Larochelle, J. (1998) Comments on a negative appraisal of taxidermic mounts as tools for studies of ecological energetics. *Physiological Zoology* **71**, 596–598.

Lathe, R. & Morris, R.G.M. (1994) Analysing brain function and dysfunction in transgenic animals. *Neuropathological Applied Neurobiology* **20**, 350–358.

Laudien, H., ed. (1986) *Temperature Relations in Animals and Man*. Fischer, Stuttgart.

Laughlin, R.B. Jr & Neff, J.M. (1981) The ontogeny of the respiratory and growth response of larval mud crabs, *Rhithropanopeus harrisii* exposed to different temperatures, salinities and naphthalene concentrations. *Marine Ecology — Progress Series* **5**, 319–332.

Lawton, J.H. (1991) From physiology to population dynamics and communities. *Functional Ecology* **5**, 155–161.

Lawton, J.H. (1995) The response of insects to environmental change. In: *Insects in a Changing Environment* (eds R. Harrington & N.E. Stork), pp. 3–26, Academic Press, London.

Lawton, J.H. (1996) Patterns in ecology. *Oikos* **75**, 145–147.

Lawton, J.H. & Woodroffe, G.L. (1991) Habitat and the distribution of water voles: why are there gaps in a species' range? *Journal of Animal Ecology* **60**, 79–91.

Layne, J.R. Jr, Claussen, D.L. & Manis, M.L. (1987) Effects of acclimation temperature, season, and time of day on the critical thermal maxima and minima of the crayfish *Orconectes rusticus*. *Journal of Thermal Biology* **12**, 183–188.

Lee, K.J. & Watts, S.A. (1994) Specific activity of Na+/K+ ATPase is not altered in response to changing salinities during early development of the brine shrimp *Artemia franciscana*. *Physiological Zoology* **67**, 910–924.

Lee, R.E. & Baust, J.G. (1982a) Absence of metabolic cold adaptation and compensatory acclimation in the Antarctic fly, *Belgica antarctica*. *Journal of Insect Physiology* **28**, 725–729.

Lee, R.E. & Baust, J.G. (1982b) Respiratory metabolism of the Antarctic tick, *Ixodes uriae*. *Comparative Biochemistry and Physiology* **72A**, 167–171.

Lee, R.E. & Baust, J.G. (1987) Cold-hardiness in the Antarctic tick, *Ixodes uriae*. *Physiological Zoology* **60**, 499–506.

Lee, R.E. & Denlinger, D.L. (1991) *Insects at Low Temperature*. Chapman & Hall, New York.

Legallois, C.J.J. (1830) *Oeuvres (avec des notes de M. Pariset)*, Paris.

LeMaho, Y., Delclitte, P. & Chatonnet, J. (1976) Thermoregulation in fasting Emperor penguins under natural conditions. *American Journal of Physiology* **231**, 913–922.

Leroi, A.M., Bennett, A.F. & Lenski, R.E. (1994a) Temperature acclimation and competitive fitness: an experimental test of the beneficial acclimation assumption. *Proceedings of the National Academy of Sciences of the USA* **91**, 1917–1921.

Leroi, A.M., Lenski, R.E. & Bennett, A.F. (1994b) Evolutionary adaptation to temperature. III. Adaptation of *Escherichia coli* to a temporally varying environment. *Evolution* **48**, 1222–1229.

Lessells, C.M. (1991) The evolution of life histories. In: *Behavioural Ecology* (eds J.R. Krebs & N.B. Davies), 3rd edn, pp. 32–68. Blackwell Scientific Publications, London.

Levins, R. (1969) Some demographic and genetic consequences of environmental heterogeneity for biological control. *Bulletin of the Entomological Society of America* **15**, 237–240.

Lewis, J.R. (1964) *The Ecology of Rocky Shores*. English Universities Press, London.

Lewontin, R.C. & Birch, L.C. (1966) Hybridization as a source of variation for adaptation to new environments. *Evolution* **20**, 315–336.

Liebig, J. (1840) *Chemistry and its Application to Agriculture and Physiology*. Taylor & Walton, London.

Lighton, J.R.B., Bartholemew, G.A. & Feener, D.H. Jr (1987) Energetics of locomotion and load carriage and a model of the energy cost of foraging in the leaf-cutting ant *Atta colombica* Guer. *Physiological Zoology* **60**, 524–537.

Lin, J.J. & Somero, G.N. (1995) Temperature-dependent changes in expression of thermostable and thermolabile isozymes of cytosolic malate dehydrogenase in the eurythermal goby fish *Gillichthys mirabilis*. *Physiological Zoology* **68**, 114–128.

Lindburg, D.G. (1998) Coming in out of the cold: animal keeping in temperate zone zoos. *Zoo Biology* **17**, 51–53.

Lindquist, S. (1986) The heat–shock response. *Annual Review of Biochemistry* **55**, 1151–1191.

Lindstedt, S.L. & Jones, J.H. (1987) Symmorphosis: the concept of optimal design. In: *New Directions in Ecological Physiology* (eds M.E. Feder, A.F. Bennett, W.W. Burggren & R.B. Huey), pp. 289–309. Cambridge University Press, Cambridge.

Lindstedt, S.L., Thomas, R.G. & Leith, D.E. (1994) Does peak inspiratory flow contribute to setting VO$_2$ max? A test of symmorphosis. *Respiration Physiology* **95**, 109–118.

Lindstedt, S.L., Wells, D.J., Jones, J.H., Hoppeler, H. & Thronson, H.A. Jr (1988) Limitations to aerobic performance in mammals: interaction of structure and demand. *International Journal of Sports Medicine* **9**, 210–217.

Ling, L., Olson, E.B. Jr, Vidrut, E.H. & Mitchell, G.S. (1997) Developmental plasticity of the hypoxic respiratory response. *Respiration Physiology* **110**, 261–268.

Lobel, P.B., Belkhode, S.P., Bajdik, C., Jackson, S.E. & Longerich, H.P. (1992) General characteristics of the frequency distributions of element concentrations and of interelemental correlations in aquatic organisms. *Marine Environmental Research* **33**, 111–126.

Lockwood, A.P.M. & Inman, C.B.E. (1973) The blood volume of some amphipod crustaceans in relation to the salinity they inhabit. *Comparative Biochemistry and Physiology* **44A**, 935–941.

Lomolino, M.V. & Channell, R. (1998) Range collapse, re-introductions, and biogeographic guidelines for conservation. *Conservation Biology* **12**, 481–484.

Loosanoff, V.L. & Nomejko, C.A. (1957) Existence of physiologically different races of oysters *Crassostrea virginica*. *Biological Bulletin* **101**, 151–156.

Losos, J.B. (1994) An approach to the analysis of comparative data when a phylogeny is unavailable or incomplete. *Systematic Biology* **43**, 117–123.

Lucas, A. (1991) Programming by early nutrition in man. In: *The Childhood Environment and Adult Disease* (eds G.R. Bock & J. Whelan), pp. 38–55. John Wiley, Chichester.

Lutz, P.L. & Storey, K.B. (1997) Adaptations to variations in oxygen tension by vertebrates and invertebrates. In: *Handbook of Physiology, Sect. 13: Comparative Physiology,* Vol 2 (ed. W.H. Dantzler), pp. 1479–1522. Oxford University Press, New York.

MacArthur, R.H. (1972) *Geographical Ecology: Patterns in the Distributions of Species.* Harper & Row, New York.

MacDonald, A.G. (1975) *Physiological Aspects of Deep Sea Biology.* Cambridge University Press, Cambridge.

MacDonald, A.G. & Gilchrist, I. (1978) Further studies on the pressure tolerance of deep sea Crustacea with observations using a high pressure trap. *Marine Biology* **45**, 9–21.

MacDonald, A.G. & Gilchrist, I. (1980) Effects of hydraulic decompression and compression of deep sea amphipods. *Comparative Biochemistry and Physiology* **67A**, 149–153.

MacDonald, A.G. & Gilchrist, I. (1982) The pressure tolerance of deep sea amphipods collected at their ambient high pressure. *Comparative Biochemistry and Physiology* **71A**, 349–352.

MacDonald, A.G., Gilchrist, I. & Teal, J.M. (1972) Some observations on the tolerance of oceanic plankton to high hydrostatic pressure. *Journal of the Marine Biological Association of the UK* **52**, 213–223.

MacDonald, A.G., Grossman, A., Marquis, R.E. *et al.* (1993) *Effects of High Pressure on Biological Systems.* Springer-Verlag, New York.

MacDonald, A.G. & Teal, J.M. (1975) Tolerance of oceanic and shallow water Crustacea to high hydrostatic pressure. *Deep-Sea Research* **22**, 131–144.

Mager, M., Blatt, W.E., Natale, P.J. & Blatteis, C.M. (1967) Influence of simulated altitude on the distribution of lactic dehydrogenase (LDH) isozymes in the neonatal and adult rat. *Federation Proceedings* **26**, 665.

Malik, M. & Camm, A.J., eds. (1995) *Heart Rate Variability.* Blackwell Science, Oxford.

Mallet, J. (1989) The evolution of insecticide resistance: have the insects won? *Trends in Ecology and Evolution* **4**, 336–340.

Maltby, L. (in press) Studying stress: the importance of organism-level responses. *Ecological Applications.*

Maltby, L., Snart, J.O.H. & Calow, P. (1987) Acute toxicity tests in the freshwater isopod, *Asellus aquaticus,* using ferrous sulfate heptahydrate with special reference to techniques and the possibility of intraspecific variation. *Environmental Pollution* **43**, 271–280.

Mangum, C.P. (1963) Studies on the speciation in malanid polychaetes of the North American Atlantic coast: III. Intraspecific and interspecific divergence in oxygen consumption. *Comparative Biochemistry and Physiology* **10**, 335–349.

Mangum, C.P. (1972) Temperature sensitivity of metabolism in offshore and intertidal oriuphid polychaetes. *Marine Biology* **17**, 108–114.

Mangum, C.P. (1978) Temperature adaptation. In: *Physiology of Annelids* (ed. P.J. Mill), pp. 447–478. Academic Press, London & New York.

Mangum, C.P. (1997) Invertebrate blood oxygen carriers. In: *Handbook of Physiology, Section 13: Comparative Physiology,* Vol 2 (ed. W.H. Dantzler), pp. 1097–1135. Oxford University Press, New York.

Mangum, C.P. & Hochachka, P.W. (1998) New directions in comparative physiology and biochemistry. Mechanisms, adaptations and evolution. *Physiological Zoology* **71**, 471–484.

Mangum, C.P. & Towle, D.W. (1977) Physiological adaptation to unstable environments. *American Science* **65**, 67–75.

Mani, M.S. (1990) *Fundamentals of High Altitude Biology,* 2nd edn. Aspect Publications, London.

Manwell, C., Baker, C.M.A. & Childers, W. (1963) The genetics of hemoglobin in hybrids. I. A molecular basis for hybrid vigor. *Comparative Biochemistry and Physiology* **10**, 103–120.

Markel, R.P. (1974) Aspects of the physiology of temperature acclimation in the limpet *Acmaea limatula* Carpenter (1864): an integrated field and laboratory study. *Physiological Zoology* **47**, 99–109.

Markert, C.L. (1963) Epigenetic control of specific protein synthesis in differentiating cells. In: *Cytodifferentiation and Macromolecular Synthesis* (ed. M. Locke), pp. 65–84. Academic Press, New York.

Maroni, G.J., Wise, J., Young, J.E. & Otto, E.

(1987) Metallothionein gene duplications and metal tolerance in natural populations of *Drosophila melanogaster*. *Genetics* **117**, 739–744.

Martínez, D.E. & Levington, J. (1996) Adaptation to heavy metals in the aquatic oligochaete *Limnodrilus hoffmeisteri*: evidence for control by one gene. *Evolution* **50**, 1339–1343.

Mason, R.P., Mangum, C.P. & Godette, G. (1983) The influence of organic ions and acclimation salinity on hemocyanin–oxygen binding in the blue crab *Callinectes sapidus*. *Biological Bulletin* **164**, 104–123.

Massion, D.D. (1983) An altitudinal comparison of water and metabolic relations in two acridid grasshoppers (Orthoptera). *Comparative Biochemistry and Physiology* **74A**, 101–105.

Matthews, W.J. (1986) Geographic variation in thermal tolerance of a widespread minnow *Notropis lutrensis* of the North American mid-west. *Journal of Fish Biology* **28**, 407–417.

Maurer, B.A., Brown, J.H. & Rusler, R.D. (1992) The micro and macro in body size evolution. *Evolution* **46**, 939–953.

May, R.M. (1975) Patterns of species abundance and diversity. In: *Ecology and Evolution of Communities* (eds M.L. Cody & J.M. Diamond), pp. 81–120. Harvard University Press, Cambridge, Massachusetts.

May, R.M. (1978) The dynamics and diversity of insect faunas. In: *Diversity of Insect Faunas* (eds L.A. Mound & N. Waloff), pp. 188–204. Blackwell Scientific Publications, Oxford.

May, R.M. (1989) Levels of organisation in ecology. In: *Ecological Concepts* (ed. J.M. Cherrett), pp. 339–363. Blackwell Scientific Publications, Oxford.

Mayer, A.G. (1914) The effect of temperature on tropical marine animals. *Papers of the Tortugas Marine Laboratory* **6**, 1–24.

Maynard Smith, J. (1989) *Evolutionary Genetics*. Oxford University Press, Oxford.

Maynard Smith, J., Burian, R., Kauffman, S., Alberch, P., Campbell, J., Goodwin, B., Lande, R., Ravp, D. & Wolpert, L. (1985) Developmental constraints and evolution. *Quarterly Review of Biology* **60**, 265–287.

Mayr, E. (1963) *Animal Species and Evolution*. Harvard University Press, Massachusetts.

McArdle, B.H., Gaston, K.J. & Lawton, J.H. (1990) Variation in the size of animal populations: patterns, processes and artefacts. *Journal of Animal Ecology* **59**, 439–454.

McCormack, J.H., Jones, B.R. & Hokanson, K.E.F. (1977) White sucker (*Catostomus commersoni*) embryo development, early growth and survival at different temperatures. *Journal of the Fisheries Research Board of Canada* **34**, 1019–1025.

McCormick, S.D. (1994) Ontogeny and evolution of salinity tolerance in anadromous salmonids: hormones and heterochrony. *Estuaries* **17**, 26–33.

McCormick, S.D., Naiman, R.J. & Montgomery, E.T. (1985) Physiological smolt characteristics of anadromous and non-anadromous brook trout (*Salvelinus fontinalis*) and Atlantic salmon (*Salmo salar*). *Canadian Journal of Fisheries and Aquatic Sciences* **42**, 529–538.

McCormick, S.D. & Saunders, R.L. (1987) Preparatory physiological adaptations for marine life of salmonids: osmoregulation, growth and metabolism. *American Fisheries Society Symposium* **1**, 211–229.

McKinney, M.L. & McNamara, K.J. (1991) *Heterochrony: The Evolution of Ontogeny*. Plenum Press, New York.

McMahon, R.F. (1990) Thermal tolerance, evaporative water loss, air–water oxygen consumption and zonation of intertidal prosobranchs: a new synthesis. *Hydrobiologia* **193**, 241–260.

McManus, M.G., Place, A.R. & Zamer, W.E. (1997) Physiological variation among clonal genotypes in the sea anemone *Haliplanella lineata*: growth and biochemical content. *Biological Bulletin* **192**, 426–443.

McNab, B.K. (1992) A statistical analysis of mammalian rates of metabolism. *Functional Ecology* **6**, 672–679.

McNab, B.K. (1994) Energy conservation and the evolution of flightlessness in birds. *American Naturalist* **144**, 628–642.

McNamara, K.J., ed. (1995) *Evolutionary Change and Heterochrony*. John Wiley, Chichester.

McNamara, J.C., Moreira, G.S. & Moreira, P.S. (1985) Thermal effects on metabolism in selected ontogenic stages of the freshwater shrimps *Macrobrachium olfersii* and *Macrobrachium heterochirus* (Decapoda, Palaemonidae). *Comparative Biochemistry and Physiology* **80A**, 187–190.

Meadows, P.S. & Campbell, J.I. (1988) *An Introduction to Marine Science*, 2nd edn. Blackie, Glasgow and London.

Meffe, G.K., Weeks, S.C., Mulvey, M. & Kandl, K.L. (1995) Genetic differences in thermal tolerance of eastern mosquitofish (*Gambusia holbrooki*: Poeciliidae) from ambient and thermal ponds. *Canadian Journal of Fisheries and Aquatic Sciences* **52**, 2704–2711.

Mehlman, D.W. (1997) Change in avian abundance across the geographic range in response to environmental change. *Ecological Applications* **7**, 614–624.

Merriam, C.H. (1894) Laws of temperature control of the geographic distribution of terrestrial animals and plants. *National Geographic* **6**, 229–238.

Metalli, P. & Ballardin, E. (1972) Radiobiology of *Artemia*: radiation effects and ploidy.

Current Topics in Radiation Research Quarterly **7**, 181–240.

Metcalfe, N.B., Taylor, A.C. & Thorpe, J.E. (1995) Metabolic rate, social status and life history strategies in Atlantic salmon. *Animal Behavior* **49**, 431–436.

Micol, T., Doncaster, C.P. & Mackinlay, L.A. (1994) Correlates of local variation in the abundance of hedgehogs *Erinaceus europaeus*. *Journal of Animal Ecology* **63**, 851–860.

Miles, D.B. & Dunham, A.E. (1993) Historical perspectives in ecology and evolutionary biology; the use of phylogenetic comparative analyses. *Annual Review of Ecology and Systematics* **24**, 587–619.

Miller, K. & Packard, G.C. (1977) An altitude cline in critical thermal maxima of chorus frogs (*Pseudacris triseriata*). *American Naturalist* **111**, 267–277.

Mitton, J.B. (1995) Enzyme heterozygosity and developmental stability. *Acta Theriologica* **40** (Suppl.), 33–54.

Mitton, J.B., Carey, C. & Kocher, T.D. (1986) The relation of enzyme heterozygosity to standard and active oxygen consumption and body size of tiger salamanders, *Ambystoma tigrinum*. *Physiological Zoology* **59**, 574–582.

Møller, A.P. (1997) Developmental stability and fitness: a review. *American Naturalist* **149**, 916–932.

Møller, A.P. & Swaddle, J.P. (1997) *Developmental Stability and Evolution*. Oxford University Press, Oxford.

Monge, C. & Léon-Velarde, F. (1991) Physiological adaptation to high altitude oxygen transport in mammals and birds. *Physiological Reviews* **71**, 1135–1172.

Monteith, J.L. (1973) *Principles of Environmental Physics*. Edward Arnold, London.

Moore, W.G. (1942) Field studies on the oxygen requirements of certain fresh-water fishes. *Ecology* **23**, 319–329.

Moore, P.D., Chaloner, W.G. & Stott, P.A. (1996) *Global Environmental Change*. Blackwell Science, Oxford.

Moore, R.O. & Villee, C.A. (1963) Multiple molecular forms of malate dehydrogenases in echinoderm embryos. *Comparative Biochemistry and Physiology* **9**, 81–94.

Morimoto, R.I., Tissieres, A. & Georgopoulos, C., eds. (1994) *Heat Shock Proteins: Structure, Function and Regulation*. Cold Spring Harbor Laboratory Press, New York.

Morin, C.L. & Eckel, R.H. (1997) Transgenic and knockout rodents: novel insights into mechanisms of body weight regulation. *Journal of Nutrition and Biochemistry* **8**, 702–706.

Morris, S. & Taylor, A.C. (1983) Diurnal and seasonal variation in physico-chemical conditions within intertidal rock pools.

Estuarine and Coastal Shelf Science **17**, 339–355.

Morritt, D. (1987) Evaporative water loss under desiccation stress in semiterrestrial and terrestrial amphipods (Crustacea: Amphipoda: Talitridae). *Journal of Experimental Marine Biological Ecology* **111**, 145–157.

Morritt, D. & Richardson, A.M.M. (1998) Female control of the embryonic environment in a terrestrial amphipod, *Mysticotalitrus cryptus* (Crustacea). *Functional Ecology* **12**, 351–358.

Morritt, D. & Spicer, J.I. (1995) Changes in the pattern of osmoregulation in the brackish water amphipod *Gammarus duebeni* Lilljeborg (Crustacea) during embryonic development. *Journal of Experimental Zoology* **273**, 271–281.

Morritt, D. & Spicer, J.I. (1996a) Developmental ecophysiology of the beachflea *Orchestia gammarellus* (Pallas) (Crustacea: Amphipoda). II. Ontogeny of osmoregulation in embryos. *Journal of Experimental Marine Biology and Ecology* **207**, 205–216.

Morritt, D. & Spicer, J.I. (1996b) Developmental ecophysiology of the beachflea *Orchestia gammarellus* (Pallas) (Crustacea: Amphipoda). I. Female control of the embryonic environment. *Journal of Experimental Marine Biology and Ecology* **207**, 191–203.

Morritt, D. & Spicer, J.I. (1999) Developmental ecophysiology of the beachflea *Orchestia gammarellus* (Pallas) (Crustacea: Amphipoda). III. Physiological competency. *Journal of Experimental Marine Biology and Ecology* **232**, 275–283.

Mortola, J.P., Rezzonico, R. & Lanthier, C. (1989) Ventilation and oxygen consumption during acute hypoxia in newborn mammals. A comparative analysis. *Respiration Physiology* **78**, 31–43.

Moss, S. (1998) Predictions of the effects of global climate change on Britain's birds. *British Birds* **91**, 307–325.

Mossman, H.W. (1987) *Vertebrate Fetal Membranes*. Rutgers University Press, Piscataway.

Mott, J.C. (1961) Ability of young mammals to withstand total oxygen lack. *British Medical Bulletin* **17**, 144–148.

Murawski, S.A. (1993) Climate change and marine fish distributions: forecasting from historical analogy. *Transactions of the American Fisheries Society* **122**, 647–658.

Natochin, Y.V. & Chernigovskaya, T.V. (1997) Evolutionary physiology: history, principles. *Comparative Biochemistry and Physiology* **118A**, 63–79.

Nee, S., Read, A.F., Greenwood, J.J.D. & Harvey, P.H. (1991) The relationship between abundance and body size in British birds. *Nature* **351**, 312–313.

Needham, J. (1928) The developmental efficiency of the avian embryo. *Journal of Experimental Biology* **5**, 43–51.

Needham, J. (1930) The biochemical aspect of the recapitulation theory. *Biological Reviews* **5**, 142–158.

Needham, J. (1931) *Chemical Embryology* (2 volumes). Cambridge University Press, Cambridge.

Needham, J. (1938) Contributions of chemical physiology to the problem of reversibility in evolution. *Biological Reviews* **13**, 225–251.

Needham, J. (1942) *Biochemistry and Morphogenesis*. Cambridge University Press, Cambridge.

Needham, J., Brachet, J. & Brown, R. (1935) The origin and fate of urea in the developing hen's egg. *Journal of Experimental Biology* **12**, 321–336.

Nelson, J.A. (1990) Muscle metabolite response to exercise and recovery in yellow perch (*Perca flavescens*): Comparison of populations from naturally acidic and neutral waters. *Physiological Zoology* **63**, 886–908.

Nelson, J.A. & Mitchell, G.S. (1992) Blood chemistry response to acid exposure in yellow perch (*Perca flavescens*): comparison of populations from naturally acidic and neutral environments. *Physiological Zoology* **65**, 493–514.

Nelson, J.A., Tang, Y. & Boutilier, R.G. (1994) Differences in exercise physiology between two Atlantic cod (*Gadus morhua*) populations from different environments. *Physiological Zoology* **67**, 330–354.

Nelson, S.G., Armstrong, D.A., Knight, A.W. & Li, H.W. (1977) The effects of temperature and salinity on the metabolic rate of juvenile *Macrobrachium rosenbergi* (Crustacea: Palaemonidae). *Comparative Biochemistry and Physiology* **56A**, 533–537.

Nestler, J.R. (1990) Relationships between respiratory quotient and metabolic rate during entry to and arousal from daily torpor in deer mice (*Peromyscus maniculatus*). *Physiological Zoology* **63**, 504–515.

Nestler, J.R., Peterson, S.J., Smith, B.D., Heathcock, R.B., Johanson, C.R., Sarthou, J.D. & King J.C. (1997) Glycolytic enzyme binding during entrance to daily torpor in deer mice (*Peromyscus maniculatus*). *Physiological Zoology* **70**, 61–67.

Nevo, E. & Shkolnik, A. (1974) Adaptive metabolic variation of chromosomal forms in mole rats *Spalax*. *Experientia* **30**, 724–726.

Nevo, E., Simson, S., Beiles, A. & Yahav, S. (1989) Adaptive variation in structure and function of kidneys of speciating subterranean mole rats. *Oecologia* **82**, 210–216.

Newell, R.C. (1979) *Biology of Intertidal Animals*, 3rd edn. Marine Ecological Surveys, Faversham, Kent.

Newlands, H.W., McMillen, W.N. & Reineke, E.P. (1952) Temperature adaptation in the baby pig. *Journal of Animal Science* **11**, 118–133.

Newton, I. (1998) *Population Limitation in Birds*. Academic Press, London.

Newton, I., Wyllie, I. & Rothery, P. (1993) Annual survival of sparrowhawks *Accipiter nisus* breeding in three areas of Britain. *Ibis* **135**, 49–60.

Nilsson, S. & Holmgreen, S. (1994) *Comparative Physiology and Evolution of the Autonomic Nervous System*. Gordon & Breach, Newark.

Nottebohn, F. (1981) A brain for all seasons: cyclical anatomical changes in song-control nuclei of the canary brain. *Science* **214**, 1368–1370.

Nylund, L. (1991) Metabolic rates of *Calathus melanocephalus* (L.) (Coleoptera, Carabidae) from alpine and lowland habitats (Jeløy and Finse, Norway and Drenthe, The Netherlands). *Comparative Biochemistry and Physiology* **100A**, 853–862.

O'Connor, T.P. (1995) Seasonal acclimatization of lipid mobilization and catabolism in house finches (*Carpodacus mexicanus*). *Physiological Zoology* **68**, 985–1005.

von Oertzen, J.-A. (1984) Metabolic similarity of *Palaemon* populations from different brackish waters. *Internationale Revue Gesamten Hydrobiologie* **69**, 753–755.

Ohno, S. (1995) Why ontogeny recapitulates phylogeny. *Electrophoresis* **16**, 1782–1786.

Okubo, S. & Mortola, J.P. Jr (1988) Long-term respiratory effects of neonatal hypoxia in the rat. *Journal of Applied Physiology* **64**, 952–958.

Okubo, S. & Mortola, J.P. Jr (1990) Control of ventilation in adult rats hypoxic in the neonatal period. *American Journal of Physiology* **259**, R836–R841.

Onimaru, H., Ballanyi, K. & Richter, D.W. (1996) Calcium-dependent responses in neurons of the isolated respiratory network of newborn rats. *Journal of Physiology* **491**, 677–695.

Orlando, K. & Pinder, A.W. (1995) Larval cardiovascular ontogeny and allometry in *Xenopus laevis*. *Physiological Zoology* **68**, 63–75.

Ostbye, K., Oxnevad, S.A. & Vollestad, L.A. (1997) Developmental stability in perch (*Perca fluviatilis*) in acidic aluminium-rich lakes. *Canadian Journal of Zoology* **75**, 919–928.

Otis, E.M. & Brent, R. (1954) Equivalent ages in mouse and human embryos. *Anatomical Record* **120**, 33–63.

Pace, N., Meyer, L.B. & Vaughan, B.E. (1956) Erythrolysis on return of altitude acclimated individuals to sea level. *Journal of Applied Physiology* **9**, 141–144.

Packard, G.C. & Janzen, F.J. (1996) Interpopulational variation in the cold-tolerance of hatchling painted turtles. *Journal of Thermal Biology* **21**, 183–190.

Paine, R.T. (1974) Intertidal community structure. Experimental studies on the relationship between a dominant competitor and its principal predator. *Oecologia* **15**, 93–120.

Pantelouris, E.M. (1967) *Introduction to Animal Physiology and Physiological Genetics*. Pergamon Press, Oxford.

Parmesan, C. (1996) Climate and species' range. *Nature* **382**, 765–766.

Parry, G. (1960) The development of salinity tolerance in the salmon, *Salmo salar* (L.) and some related species. *Journal of Experimental Biology* **37**, 425–434.

Parsons, P.A. (1996a) Competition versus abiotic factors in variably stressful environments: evolutionary implications. *Oikos* **75**, 129–132.

Parsons, P.A. (1996b) Conservation strategies: adaptation to stress and the preservation of genetic diversity. *Biological Journal of the Linnaean Society* **58**, 471–482.

Paton, J.F.R., Ramirez, J.M. & Richter, D.W. (1994) Mechanisms of respiratory rhythm generation change profoundly during early life in mice and rats. *Neuroscience Letters* **170**, 167–170.

Pattee, E., Lascombe, C. & Delolme, R. (1973) Effects of temperature on the distribution of turbellarian triclads. In: *Effects of Temperature on Ectothermic Organisms* (ed. W. Wieser), pp. 201–208. Springer-Verlag, Berlin.

Patterson, J.W. (1984) Thermal acclimation in two subspecies of the tropical lizard *Mabuya striata*. *Physiological Zoology* **57**, 301–306.

Peach, W.J., Baillie, S.R. & Underhill, L. (1991) Survival of British sedge warblers *Acrocephalus schoenobaenus* in relation to West African rainfall. *Ibis* **133**, 300–305.

Pearse, A.S. (1923) *Animal Ecology*. McGraw-Hill Book Company, New York.

Pearse, A.S. (1939) *Animal Ecology*, 2nd edn. McGraw-Hill Book Company, New York.

Pearson, O.P. (1954) The daily energy requirements of a wild Anna hummingbird. *Condor* **56**, 317–322.

Pearson, J.T., Tsudzuki, M., Nakane, Y., Pikiyama, R. & Tazawa, H. (1988) Development of heart rate in the precocial King Quail *Coturnix chinensis*. *Journal of Experimental Biology* **201**, 931–941.

Peiss, C.N. & Field, J. (1950) The respiratory metabolism of excised tissues of warm- and cold-adapted fishes. *Biological Bulletin* **99**, 213–224.

Pepper, J.H. (1939) The effect of certain climatic factors on the distribution of the beet webworm (*Loxostege sticticalis* L.) in North America. *Ecology* **19**, 565–571.

Perutz, M.F. (1983) Species adaptation in a protein molecule. *Molecular Biology and Evolution* **1**, 1–28.

Peters, R.H. (1983) *The Ecological Implications of Body Size*. Cambridge University Press, New York.

Peters, R.L. & Lovejoy, T.E., eds. (1992) *Global Warming and Biological Diversity*. Yale University Press, New Haven.

Peters, J.J., Vonderahe, A.R. & Huesman, A.A. (1960) Chronological development of electrical activity in the optic lobes, cerebellum and cerebrum of the chick embryo. *Physiological Zoology* **33**, 225–231.

Peterson, C.C., Nagy, K.A. & Diamond, J. (1990) Sustained metabolic scope. *Proceedings of the National Academy of Sciences of the USA* **87**, 2324–2328.

Petschow, D., Würdunger, I., Baumann, R. *et al.* (1977) Causes of high blood O_2 affinity of animals living at high altitudes. *Journal of Applied Physiology* **42**, 139–143.

Phillipson, J. (1981) Bioenergetic options and phylogeny. In: *Physiological Ecology: An Evolutionary Approach to Resource Use* (eds C.R. Townsend & P. Calow), pp. 20–45. Blackwell Scientific Publications, Oxford.

Pianka, E.R. (1966) Latitudinal gradients in species diversity: a review of concepts. *American Naturalist* **100**, 33–46.

Pickens, P.E. (1965) Heart rate of mussels as a function of latitude, intertidal height and acclimation temperature. *Physiological Zoology* **38**, 390–405.

Pierce, V.A. & Crawford, D.L. (1996) Variation in the glycolytic pathway: the role of evolutionary and physiological processes. *Physiological Zoology* **69**, 489–508.

Piersma, T., Bruinzeel, L., Drent, R., Kersten, M., Van der Meer, J. & Wiersma, P. (1996) Variability in basal metabolic rate of a long-distance migrant shorebird (red knot, *Calidris canutus*) reflects shifts in organ sizes. *Physiological Zoology* **69**, 191–217.

Piiper, J., ed. (1978) *Respiratory Function in Birds, Adult and Embryonic*. Springer-Verlag, Berlin.

Pinder, A.W. & Friet, S.C. (1994) Oxygen transport in egg masses of the amphibians *Rana sylvatica* and *Ambystoma maculatum*: convection, diffusion and oxygen production by algae. *Journal of Experimental Biology* **197**, 17–30.

Place, A.R. & Powers, D.A. (1979) Genetic variation and relative catalytic efficiencies: lactate dehydrogenase B allozymes of *Fundulus heteroclitus*. *Proceedings of the National Academy of Sciences of the USA* **76**, 2354–2358.

Pohl, H. (1976) Seasonal changes in the whole animal. In: *Environmental Physiology of Animals* (eds J. Bligh, J.L. Cloudsley-Thompson & A.G. Macdonald), pp. 311–336. Blackwell Scientific Publications, Oxford.

Pollard, E. (1988) Temperature, rainfall and butterfly numbers. *Journal of Applied Ecology* **25**, 819–828.

Ponat, A. (1967) Untersuchungen zur zellulären Druckresistenz Verschiedener Everterbraten der Nord- und Ostee. *Kieler Sonderh Meeresforsch* **23**, 21–47.

Pörtner, H.O. & Playle, R., eds. (1998) *Cold Ocean Physiology*. Cambridge University Press, Cambridge.

Posthuma, L., Hogervorst, R.F., Joose, E.N.G. & Van Straalen, N.M. (1993) Genetic variation and covariation for characteristics associated with cadmium tolerance in natural populations of the springtail *Orchesella cincta* (L.). *Evolution* **47**, 619–631.

Potts, W.T.W. & Parry, G. (1964) *Osmotic and Ionic Regulation in Animals*. Pergamon Press, Oxford.

Powers, D.A. (1987) A multidisciplinary approach to the study of genetic variation within species. In: *New Directions in Ecological Physiology* (eds M.E. Feder, A.F. Bennett, W.W. Burggren & R.B. Huey), pp. 102–129. Cambridge University Press, Cambridge.

Powers, D.A., Ropson, I., Brown, D.C., Van Beneden, R., Cashon, R., Gonzalez-Villasenor, L.I. & Di Michele, L. (1986) Genetic variation in *Fundulus heteroclitus*: geographic distribution. *American Zoologist* **26**, 131–144.

Prats, M.T., Palacios, L., Gallego, S. & Riera, M. (1996) Blood oxygen transport properties during migration to higher altitude of wild quail, *Coturnix coturnix coturnix*. *Physiological Zoology* **69**, 912–929.

Praud, J.-P., Diaz, V., Kianicka, I., Chevalier, J.Y., Canet, E. & Thisdale, Y. (1997) Abolition of breathing rhythmicity in lambs by CO_2 unloading in the first few hours of life. *Respiration Physiology* **110**, 1–8.

Precht, H. (1958) Concepts of the temperature adaptation of unchanging reaction systems of cold-blooded animals. In: *Physiological Adaptation* (ed. C.L. Prosser), pp. 50–78. American Physiological Society, Washington, DC.

Precht, H. (1973) Limiting temperatures of life functions. In: *Temperature and Life* (eds H. Precht, J. Christophersen, H. Hensel & W. Larcher), pp. 400–440. Springer-Verlag, Berlin.

Precht, H., Christophersen, J., Hensel, H. & Larcher, W., eds. (1973) *Temperature and Life*. Springer-Verlag, Berlin.

Precht, H., Christopherson, J. & Hensel, H. (1955) *Temperatur und Leben*. Springer-Verlag, Berlin.

Preston, F.W. (1948) The commonness, and rarity, of species. *Ecology* **29**, 254–283.

Preuss, T., Lebaric, Z.N. & Gilly, W.F. (1997) Post-hatching development of circular mantle muscles in the squid *Loligo opalescens*. *Biological Bulletin* **192**, 375–387.

Priede, I.G. & Tytler, P. (1977) Heart rate as a measure of metabolic rate in teleost fishes; *Salmo gairdneri, Salmo trutta* and *Gadus morhua. Journal of Fish Biology* **18**, 231–242.

Primack, R.B. (1993) *Essentials of Conservation Biology*. Sinauer Associates, Sunderland.

Primmett, D.R.N., Eddy, F.B., Miles, M.S., Talbot, C. & Thorpe, J.E. (1988) Transepithelial ion exchange in smolting Atlantic salmon (*Salmo salar* L.). *Fish Physiological Biochemistry* **5**, 181–186.

Promislow, D.E.L. (1991) The evolution of mammalian blood parameters: patterns and their interpretation. *Physiological Zoology* **64**, 393–431.

Prosser, C.L. (1955) Physiological variation in animals. *Biological Reviews* **30**, 229–262.

Prosser, C.L. (1957a) Proposal for the study of physiological variation in marine animals. *Année Biologie* **33**, 191–197.

Prosser, C.L. (1957b) The species problem from the viewpoint of a physiologist. In: *The Species Problem* (ed. E. Mayr), pp. 167–180. American Association for the Advancement of Science, Washington, DC.

Prosser, C.L. (1958) General summary: The nature of physiological adaptation. In: *Physiological Adaptation* (ed. C.L. Prosser), pp. 167–180. American Physiological Society, Washington, DC.

Prosser, C.L. (1960) Comparative physiology in relation to evolutionary theory. In: *Evolution After Darwin*, Vol. 1 (ed. S. Tax), pp. 569–594. University of Chicago Press, Chicago.

Prosser, C.L. (1964a) Comparative physiology and biochemistry: status and prospects. *Comparative Biochemistry and Physiology* **11**, 1–7.

Prosser, C.L. (1964b) Perspectives of adaptation: theoretic aspects. In: *Handbook of Physiology, Sect. 4* (ed. D.B. Dill), pp. 11–25. American Physiological Society, Washington, DC.

Prosser, C.L. (1965) Levels of biological organisation and their physiological significance. In: *Ideas in Modern Biology* (ed. J.A. Moore), *Proceedings of the 16th International Congress in Zoology* **6**, 357–390. Natural History Press, New York.

Prosser, C.L. (1975) Prospects for comparative physiology and biochemistry. *Journal of Experimental Zoology* **194**, 345–348.

Prosser, C.L. (1986) *Adaptational Biology: Molecules to Organisms*. John Wiley, New York.

Prosser, C.L. (1991) Introduction: definition of comparative physiology: Theory of adaptation. In: *Comparative Animal Physiology: Environmental and Metabolic Animal Physiology*, 4th edn (ed. C.L. Prosser), pp. 1–12. Wiley-Liss, New York.

Prosser, C.L., ed. (1950) *Comparative Animal Physiology*, 1st edn. W.B. Saunders, Philadelphia and London.

Prosser, C.L., ed. (1973) *Comparative Animal Physiology*, 3rd edn. W.B. Saunders, Philadelphia and London.

Prosser, C.L. & Brown, F.A. Jr (1961) *Comparative Animal Physiology*, 2nd edn. W.B. Saunders, Philadelphia and London.

Pruett, S.J., Hoyt, D.F. & Stiffer, D.F. (1991) The allometry of osmotic and ionic regulation in Amphibia with emphasis on intraspecific scaling in larval *Ambystoma tigrinum*. *Physiological Zoology* **64**, 1173–1199.

Pugh, F.H. (1989) Organismal performance and Darwinian fitness: approaches and interpretations. *Physiological Zoology* **62**, 199–236.

Purvis, A. & Garland, T. Jr (1993) Polytomies in comparative analyses of continuous characters. *Systematic Biology* **42**, 569–575.

Purvis, A. & Rambaut, A. (1995) Comparative analysis by independent contrasts (CAIC): an Apple MacIntosh application for analysing comparative data. *Computer Applications in the Biosciences* **11**, 247–251.

Rahbek, C. (1995) The elevational gradient of species richness: a uniform pattern? *Ecography* **18**, 200–205.

Rainbow, P.S. (1997) Trace metal accumulation in marine invertebrates: marine biology or marine chemistry? *Journal of the Marine Biological Association of the UK* **77**, 195–210.

Randall, D.J., Burggren, W.W. & French, K. (1997) *Eckert Animal Physiology*, 4th edn. W.H. Freeman, New York.

Rankin, J.C. & Davenport, J. (1981) *Animal Osmoregulation*. John Wiley, New York.

Rao, K.P. (1953) Rate of water propulsion in *Mytilus californianus* as a function of latitude. *Biological Bulletin* **104**, 171–181.

Rao, K.P. & Bullock, T.H. (1954) Q_{10} as a function of size and habitat temperature in poikilotherms. *American Naturalist* **88**, 33–44.

Rapola, J. & Koskimies, O. (1967) Embryonic enzyme patterns: characterization of the single lactic dehydrogenase isozyme in preimplanted mouse ova. *Science* **157**, 1311–1312.

Rapoport, E.H. (1982) *Areography: Geographical Strategies of Species*. Pergamon, Oxford.

Ratterman, R.J. & Ackerman, R.A. (1989) The water exchange and hydric microclimate of painted turtle (*Chrysemys picta*) eggs incubating in field nests. *Physiological Zoology* **62**, 1059–1079.

Read, G.H.L. (1984) Intraspecific variation in the osmoregulatory capacity of larval, postlarval, juvenile and adult *Macrobrachium petersi*. *Comparative Biochemistry and Physiology* **78A**, 501–506.

Read, A.F. & Harvey, P.H. (1989) Life history differences among the eutherian radiations. *Journal of Zoology* **219**, 329–353.

Refinetti, R. (1996) The body temperature rhythm of the thirteen-lined ground squirrel, *Spermophilus tridecemlineatus*. *Physiological Zoology* **69**, 270–275.

Regal, P.J. (1977) Evolutionary loss of useless features: is it molecular noise suppression? *American Naturalist* **11**, 123–133.

Reilly, S.M., Wiley, E.O. & Meinhardt, D.J. (1997) An integrative approach to heterochrony: the distinction between interspecific and intraspecific phenomena. *Biological Journal of the Linnaean Society* **60**, 119–143.

Reiss, M. (1931) Das Verhalten des Stoffwechsels bei der Erstickung neugeborener Ratten und Mäuse. *Zeitschrift für Gesamte Experimentelle Medizin* **79**, 345–359.

Repasky, R.R. (1991) Temperature and the northern distributions of wintering birds. *Ecology* **72**, 2274–2285.

Reynolds, S.M.R. (1949) Perspectives in prematurity. *American Journal of Obstetrics and Gynecology* **58**, 65–74.

Richardson, M.K., Hanken, J., Gooneratne, M.L., Pieau, C., Reynaud, A., Selwood, L. & Wright, G.M. (1997) There is no highly conserved embryonic stage in the vertebrates: implications for current theories of evolution and development. *Anatomical Embryology* **196**, 91–106.

Richter, T.A., Webb, P.I. & Skinner, J.D. (1997) Limits to the distribution of the southern African Ice Rat (*Otomys sloggetti*): thermal physiology or competitive exclusion? *Functional Ecology* **11**, 240–246.

Ricketts, E.F. & Calvin, J. (1968) *Between Pacific Tides*, 4th edn. Stanford University Press, Stanford, California.

Roberts, J.L. (1957) Thermal acclimation of metabolism in the crab *Pachygrapsus crassipes* Randall. II. Mechanisms and the influence of season and latitude. *Physiological Zoology* **30**, 242–255.

Roberts, D.A., Hofmann, G.E. & Somero, G.N. (1997) Heat-shock protein expression in *Mytilus californianus*: acclimatization (seasonal and tidal-height comparisons) and acclimation effects. *Biological Bulletin* **192**, 309–320.

Roberts, T.J., Weber, J.M., Hoppeler, H., Weibel, E.R. & Taylor, C.R. (1996) Design of the oxygen and substrate pathways. II. Defining the upper limits of carbohydrate and fat oxidation. *Journal of Experimental Biology* **199**, 1651–1658.

Robinson, T., Rogers, D. & Williams, B. (1997a) Univariate analysis of tsetse habitat in the common fly belt of Southern Africa using climate and remotely sensed vegetation data. *Medical and Veterinary Entomology* **11**, 223–234.

Robinson, T., Rogers, D. & Williams, B. (1997b) Mapping tsetse habitat suitability in the common fly belt of Southern Africa using multivariate analysis of climate and remotely sensed vegetation data. *Medical and Veterinary Entomology* **11**, 235–245.

Rogers, C.G. (1938) *Textbook of Comparative Physiology*, 2nd edn. McGraw-Hill, New York and London.

Rogers, D.J. & Williams, B.G. (1994) Tsetse distribution in Africa: seeing the wood *and* the trees. In: *Large-Scale Ecology and Conservation Biology* (eds P.J. Edwards, R.M. May & N.R. Webb), pp. 247–271. Blackwell Scientific Publications, Oxford.

Rogowitz, G.L. (1992) Postnatal variation in metabolism and body temperature in a precocial lagomorph. *Functional Ecology* **6**, 666–671.

Rohde, K. (1992) Latitudinal gradients in species diversity: the search for the primary cause. *Oikos* **65**, 514–527.

Rollo, C.D. (1994) *Phenotypes: Their Epigenetics, Ecology and Evolution*. Chapman & Hall, London.

Romanoff, A.L. (1967) *Biochemistry of the Avian Embryo*. John Wiley, New York.

Rombough, P.J. (1988a) Respiratory gas exchange, aerobic metabolism, and effects of hypoxia during early life. In: *Fish Physiology* (eds W.S. Hoar & D.J. Randall), Vol XIA, pp. 59–161. Academic Press, New York.

Rombough, P.J. (1988b) Growth, aerobic metabolism and dissolved oxygen requirements of embryos and alevins of the steelhead trout *Salmo gairdneri*. *Canadian Journal of Zoology* **66**, 651–660.

Rombough, P.J. (1996) The effects of temperature on embryonic and larval development. In: *Global Warming: Implications for Freshwater and Marine Fish* (eds C.M. Wood & D.G. McDonald), pp. 177–223. Cambridge University Press, Cambridge.

Rombough, P.J. & Ure, D. (1991) Partitioning of oxygen uptake between cutaneous and branchial surfaces in larval and young juvenile chinook salmon *Oncorhynchus tshawytscha*. *Physiological Zoology* **64**, 717–727.

Rome, L.C., Stevens, E.D. & John-Alder, H.B. (1992) The influence of temperature and thermal acclimation on physiological function. In: *Environmental Physiology of the Amphibians* (eds M.E. Feder & W.W. Burggren), pp. 183–205. University of Chicago Press, Chicago.

Romero, J. & Real, R. (1996) Macroenvironmental factors as ultimate determinants of distribution of common toad and natterjack toad in the south of Spain. *Ecography* **19**, 305–312.

Root, T. (1988a) *Atlas of Wintering North American Birds: an Analysis of Christmas Bird Count Data*. University of Chicago Press, Chicago.

Root, T. (1988b) Environmental factors associated with avian distributional boundaries. *Journal of Biogeography* **15**, 489–505.

Root, T. (1988c) Energy constraints on avian distributions and abundances. *Ecology* **69**, 330–339.

Root, T. (1989) Energy constraints on avian distributions: a reply to Castro. *Ecology* **70**, 1183–1185.

Rosenberg, N.J. (1983) *Microclimate: The Biological Environment*, 2nd edn. Wiley-Interscience, New York.

Rosenberg, R. & Costlow, J.D. (1976) Synergistic effects of cadmium and salinity combined with constant and cycling temperatures on the larval development of two estuarine crab species. *Marine Biology* **38**, 291–303.

Rosenberg, R. & Costlow, J.D. (1979) Delayed response to irreversible non-genetic adaptation to salinity in early development of the brachyuran crab *Rhithropanopeus harrisii* and some notes on adaptation to temperature. *Ophelia* **18**, 97–112.

Rosenberg, R., Elmgren, R., Flescher, S., Jonsson, P., Persson, G. & Dahlin, H. (1990) Marine eutrophication case studies in Sweden. *Ambio* **19**, 102–108.

Rosenthal, R. (1979) The 'file drawer problem' and tolerance for null results. *Psychological Bulletin* **86**, 638–641.

Rosenthal, H. & Alderdice, D.F. (1976) Sublethal effects of environmental stressors, natural and pollutional, on marine fish eggs and larva. *Journal of the Fisheries Research Board of Canada* **33**, 2047–2065.

Rosenzweig, M.L. (1992) Species diversity gradients: we know more and less than we thought. *Journal of Mammalogy* **73**, 715–730.

Rosenzweig, M.L. (1995) *Species Diversity in Space and Time*. Cambridge University Press, Cambridge.

Roskaft, E., Jarvi, T., Bakken, M., Bech, C. & Reinertsen, R.E. (1986) The relationship between social status and resting metabolic rate in great tits (*Parus major*) and pied flycatchers (*Ficedula hypoleuca*). *Animal Behavior* **34**, 838–842.

Roverud, R.C. & Chappell, M.A. (1991) Energetic and thermoregulatory aspects of clustering behavior in the neotropical bat *Noctilio albiventris*. *Physiological Zoology* **64**, 1527–1541.

Rowe, J.S. (1961) The level-of-integration concept and ecology. *Ecology* **42**, 420–427.

Ruben, J. (1995) The evolution of endothermy

in mammals and birds: from physiology to fossils. *Annual Review of Physiology* **57**, 69–95.

Rubner, M. (1883) Über den Einfluß der Körpergrösse auf Stoff- und Kraftwechsel. *Zeitschrift für Biologie* **19**, 535–562.

Ryrholm, N. (1988) An extralimital population in a warm climatic outpost: the case of the moth *Idaea dilutaria* in Scandinavia. *International Journal of Biometeorology* **32**, 205–216.

Ryrholm, N. (1989) The influence of the climatic energy balance on living conditions and distribution patterns of *Idaea* spp. (Lepidoptera: Geometridae): an expansion of the species energy theory. *Acta Universitatis Upsaliensis* **208**.

Scadding, S.R. (1977) Phylogenetic distribution of limb regeneration capacity in adult amphibia. *Journal of Experimental Zoology* **202**, 57–68.

Scharloo, W. (1991) Canalization: genetic and developmental aspects. *Annual Review of Ecology and Systematics* **22**, 65–93.

Scharold, J. & Gruber, S.H. (1991) Telemetred heart rate as a measure of metabolic rate in the lemon shark *Negaprion brevirostris*. *Copeia* **1991**, 942–953.

Scharrer, R., Liebich, H.G., Raab, W. & Promberger, N. (1979) Influence of age and rumen development on intestinal absorption of galactose and glucose in lambs. *Zentralblatt für Veterinarmedizin Reihe* **26A**, 95–105.

Schechter, V. (1943) Tolerance of the snail *Thais floridana* to waters of low salinity and the effect of size. *Ecology* **24**, 493–499.

Scheer, B.T. (1948) *Comparative Physiology*. John Wiley, New York.

Scheiner, S.M., Caplan, R.L. & Lyman, R.F. (1991) The genetics of phenotypic plasticity. III. Plasticities and fluctuating asymmetries. *Journal of Evolutionary Biology* **4**, 51–68.

Schlieper, C. (1929) Über die Einwirkung niederer Salzkonzentrationen auf marine Organismen. *Zeitschrift für Vergleicherde Physiologie* **9**, 478–514.

Schlieper, C. (1957) Comparative study of *Asterias rubens* and *Mytilus edulis* from the North Sea (30‰) and the western Baltic Sea (15‰). *Année Biologique* **33**, 117–127.

Schlieper, C. (1960) Genotypische und phaenotypische Temperatur- und Salzgehalts-Adaptationen in Bodenevertebraten der Nord- und Ostsee. *Kieler Meeresforsch* **16**, 180–185.

Schlieper, C. (1967) Genetic and non-genetic cellular resistance adaptation in marine invertebrates. *Helgolander Wissenschaftliche Meeresuntersuchungen* **14**, 482–499.

Schlieper, C. (1968) High pressure effects on marine invertebrates and fishes. *Marine Biology* **2**, 5–12.

Schlieper, C. (1972) Comparative investigations on the pressure tolerance of marine invertebrates and fishes. *Symposium of the Society of Experimental Biology* **26**, 197–207.

Schmidt-Nielsen, K. (1984) *Scaling: Why Is Animal Size So Important?* Cambridge University Press, New York.

Schmidt-Nielsen, K. (1997) *Animal Physiology*, 5th edn. Cambridge University Press, Cambridge.

Schneider, S.H. & Root, T.L. (1996) Ecological implications of climatic change will include surprises. *Biodiversity and Conservation* **5**, 1109–1119.

Scholander, P.F., Flagg, W., Walters, V. & Irving, L. (1953) Climatic adaptation in Arctic and tropical poikilotherms. *Physiological Zoology* **26**, 67–92.

Scholander, P.F., Hock, R., Walters, V. & Irving, L. (1950a) Adaptation to cold in arctic and tropical mammals and birds in relation to body temperature, insulation and basal metabolic rate. *Biological Bulletin* **99**, 259–271.

Scholander, P.F., Hock, R., Walters, V., Johnson, F. & Irving, L. (1950b) Heat regulation in some arctic and tropical mammals and birds. *Biological Bulletin* **99**, 237–258.

Scholander, P.F., Walters, V., Hock, R. & Irving, L. (1950c) Body insulation of some arctic and tropical mammals and birds. *Biological Bulletin* **99**, 225–236.

Schuen, J.N., Bamford, O.S. & Carroll, J.N. (1997) The cardiorespiratory response to anoxia: normal development and the effect of nicotine. *Respiration Physiology* **109**, 231–239.

Schultz, T.D., Quinlan, M.C. & Hadley, N.F. (1992) Preferred body temperature, metabolic physiology, and water balance of adult *Cicindela longilabris*: a comparison of populations from boreal habitats and climatic refugia. *Physiological Zoology* **65**, 226–242.

Scott, I., Mitchell, P.I. & Evans, P.R. (1996) How does variation in body composition affect the basal metabolic rate of birds? *Functional Ecology* **10**, 307–313.

Scott, M., Berrigan, D. & Hoffmann, A.A. (1997) Costs and benefits of acclimation to elevated temperature in *Trichogramma carverae*. *Entomologica Experimentalis et Applicata* **85**, 211–219.

Secor, S.M. & Diamond, J. (1997) Determinants of the postfeeding metabolic response of Burmese pythons, *Python molurus*. *Physiological Zoology* **70**, 202–212.

Secor, S.M. & Diamond, J. (1998) A vertebrate model of extreme physiological variation. *Nature* **395**, 659–662.

Seddon, W.L. & Prosser, C.L. (1997) Seasonal variations in the temperature acclimation response of the channel catfish, *Ictalurus punctatus*. *Physiological Zoology* **70**, 33–44.

Seed, R. (1971) A physiological and biochemical approach to the taxonomy of *Mytilus edulis* (L.) and *M. galloprovincialis* (Lmk). from S.W. England. *Cahiers de Biologie Marine* **12**, 291–322.

Segal, E. (1956) Microgeographic variation as thermal acclimation in an intertidal mollusc. *Biological Bulletin* **111**, 129–152.

Seibel, B.A., Thuesen, E.V., Childress, J.J. & Gorodozky, L.A. (1997) Decline in pelagic cephalopod metabolism with habitat depth reflects differences in locomotory efficiency. *Biological Bulletin* **192**, 262–279.

Seki, K. & Toyoshima, M. (1998) Preserving tardigrades under pressure. *Nature* **395**, 853–854.

Senar, J.C. & Copete, J.L. (1995) Mediterranean house sparrows (*Passer domesticus*) are not used to freezing temperatures: an analysis of survival rates. *Journal of Applied Statistics* **22**, 1069–1074.

Seymour, R.S., Geiser, F. & Bradford, D.F. (1991) Gas conductance in the jelly capsule of terrestrial frog eggs correlates with embryonic stage, not metabolic demand or ambient oxygen partial pressure. *Physiological Zoology* **64**, 673–687.

Sheard, P.W. (1992) Physiological properties of isolated motor units in normal and bilaterally innervated *Xenopus* gastrocnemius muscles. *Developmental Brain Research* **69**, 67–75.

Shelford, V.E. (1911) Physiological animal geography. *Journal of Morphology* **22**, 551–618.

Shelford, V.E. (1930) *Laboratory and Field Ecology*. Williams and Wilkins, New York.

Shelford, V.E. & Flint, W.P. (1943) Populations of the chinch bug in the upper Mississippi valley from 1823 to 1940. *Ecology* **24**, 435–455.

Sherman, E. (1980) Ontogenic change in thermal tolerance of the toad *Bufo woodhousii fowleri*. *Comparative Biochemistry and Physiology* **65A**, 227–230.

Shick, J.M. & Dowse, H.B. (1985) Genetic basis of physiological variation in natural populations of sea anemones: intra- and interclonal analyses of variance. *Proceedings of the 19th European Marine Biology Symposium*, pp. 465–479. Cambridge University Press, Cambridge.

Shine, R. (1987) Reproductive mode may determine geographic distributions in Australian venomous snakes (*Pseudechis*, Elapidae). *Oecologia* **71**, 608–612.

Shoemaker, V.H. (1992) Exchange of water, ions and respiratory gases in terrestrial amphibians. In: *Environmental Physiology of the Amphibians* (eds M.E. Feder & W.W. Burggren), pp. 125–150. University of Chicago Press, Chicago.

Short, J., Bradshaw, S.D., Giles, J., Prince, R.I.T. & Wilson, G.R. (1992) Reintroduction of macropods (Marsupalia: Macropodoidea) in Australia: a review. *Biological Conservation* **62**, 189–204.

Sibly, R.M. & Calow, P. (1987) *Physiological Ecology of Animals: An Evolutionary Approach*. Blackwell Scientific Publications, Oxford.

Siebenaller, J.F. (1984) Analysis of the biochemical consequences of ontogenetic vertical migration in a deep-living teleost fish. *Physiological Zoology* **57**, 598–608.

Silver, M., Steven, D.H. & Comline, R.S. (1973) Placental exchange and morphology in ruminants and mare. In: *Foetal and Neonatal Physiology*, pp. 245–271. Cambridge University Press, Cambridge.

Singletary, R.L. & Shadlou, R. (1983) *Balanus balanoides* in tide-pools: a question of maladaptation? *Crustaceana* **45**, 53–70.

Skibinski, D.O.F., Beardmore, J.A. & Cross, T.F. (1983) Aspects of the population genetics of *Mytilus* (Mytilidae: Mollusca) in the British Isles. *Biological Journal of the Linnaean Society* **19**, 137–183.

Sleigh, M.A. & MacDonald, A.C., eds. (1971) The effects of pressure on organisms. *Symposium of the Society of Experimental Biology 26*. Cambridge University Press, Cambridge.

Smith, G.F.M. (1940) Factors limiting distribution and size in the starfish. *Journal of the Fisheries Research Board of Canada* **5**, 84–103.

Smith, R.F. (1954) The importance of the microenvironment in insect ecology. *Journal of Economic Entomology* **47**, 205–210.

Smith, R.I. (1955) Comparison of the level of chloride regulation by *Nereis diversicolor* in different parts of its geographical range. *Biological Bulletin* **109**, 453–474.

Smith, I.P. & Taylor, A.C. (1993) The energetic cost of agonistic behavior in the velvet swimming crab, *Necora* (= *Liocarcinus*) *puber* (L.). *Animal Behavior* **45**, 375–391.

Smith, K.L. & Hessler, R.R. (1974) Respiration of benthopelagic fishes. *In situ* measurements at 1230 meters. *Science* **184**, 72–73.

Snedecor, G.W. & Cochrane, W.G. (1989) *Statistical Methods*, 8th edn. Iowa State University Press, Ames, Iowa.

Snyder, L.R.G. (1982) 2,3-diphosphoglycerate in high- and low-altitude populations of the deer mouse. *Respiration Physiology* **48**, 107–123.

Snyder, L.R.G., Born, S. & Lechner, A.J. (1982) Blood oxygen affinity in high- and low-altitude populations of the deer mouse. *Respiration Physiology* **48**, 89–105.

Snyder, G.K. & Weathers, W.W. (1975) Temperature adaptations in amphibians. *American Naturalist* **109**, 93–101.

Soares, A.M.V.M., Baird, D.J. & Calow, P. (1992) Interclonal variation in the performance of *Daphnia magna* Staus in chronic bioassays. *Environmental Toxicology and Chemistry* **11**, 1477–1483.

Sokal, F.J. & Rohlf, R.R. (1981) *Biometry*, 2nd edn. W.H. Freeman, New York.

Somero, G.N. (1991) Biochemical mechanisms of cold-adaptation and stenothermality in Antarctic fishes. In: *Biology of Antarctic Fish* (eds G. diPrisco, B. Maresca & B. Tota), pp. 232–247. Springer-Verlag, Berlin.

Somero, G.N. (1992a) Adaptations to high hydrostatic pressure. *Annual Review of Physiology* **54**, 557–577.

Somero, G.N. (1992b) Biochemical ecology of deep-sea animals. *Experientia* **48**, 537–543.

Somero, G.N. (1995) Proteins and temperature. *Annual Review of Physiology* **57**, 43–68.

Somero, G.N. (1996) Temperature and proteins: Little things can mean a lot. *News in Physiological Sciences* **11**, 72–77.

Somero, G.N. (1997) Temperature relationships: from molecules to biogeography. In: *Handbook of Physiology, Section 13: Comparative Physiology*, Vol 2 (ed. W.H. Dantzler), pp. 1391–1444. Oxford University Press, New York.

Somero, G.N., Dahlhoff, E. & Lin, J.J. (1996) Stenotherms and eurytherms: mechanisms establishing thermal optima and tolerance ranges. In: *Animals and Temperature Phenotypic and Evolutionary Adaptation* (eds I.A. Johnston & A.F. Bennett), pp. 53–78. Cambridge University Press, Cambridge.

Somero, G.N., Siebenaller, J.F. & Hochachka, P.W. (1983) Biochemical and physiological adaptations of deep-sea animals. In: *The Sea, Ideas and Observations on Progress in the Study of the Seas*, Vol 8: *Deep-Sea Biology* (ed. G.T. Rowe), pp. 331–370. John Wiley, New York.

Sømme, L. & Block, W. (1982) Cold hardiness of Collembola at Signy Island, maritime Antarctic. *Oikos* **38**, 168–176.

Sømme, L. & Conradi-Larsen, E.M. (1977) Cold-hardiness of collembolans and oribatid mites from windswept mountain ridges. *Oikos* **29**, 118–126.

Sømme, L., Davidson, R.L. & Onore, G. (1996) Adaptation of insects at high altitudes of Chimborazo, Ecuador. *European Journal of Entomology* **93**, 313–318.

Soulier, V., Gestreau, C., Borghini, N., Dalmaz, Y., Cottetemard, J.M. & Pequiguot, J.M. (1997) Peripheral chemosensitivity and central integration: Neuroplasticity of catecholaminergic cells under hypoxia. *Comparative Biochemistry and Physiology* **118A**, 1–7.

Spanner, G. & Niessner, R. (1996) New concept of the non-invasive determination of physiological glucose concentrations using modulated laser diodes. *Fresenius' Journal of Analytical Chemistry* **354**, 306–310.

Speakman, J.R. & McQueenie, J. (1996) Limits to sustained metabolic rate: The link between food intake, basal metabolic rate, and morphology in reproducing mice *Mus musculus*. *Physiological Zoology* **69**, 746–769.

Speakman, J.R., Racey, P.A., Haim, A. *et al.* (1994) Inter- and intraindividual variation in daily energy expenditure of the pouched mouse (*Saccostomus campestris*). *Functional Ecology* **8**, 336–342.

Spector, W.S. (1956) *Handbook of Biological Data*. W.B. Saunders, Philadelphia.

Spellenberg, I.F. (1973) Critical minimum temperature of reptiles. In: *Effects of Temperature on Ectothermic Organisms* (ed. W. Wieser), pp. 239–247. Springer-Verlag, Berlin.

Spicer, J.I. (1994) Ontogeny of cardiac function in the brine shrimp *Artemia franciscana* Kellogg 1906 (Branchiopoda: Anostraca). *Journal of Experimental Zoology* **270**, 505–516.

Spicer, J.I. (1995) Ontogeny of respiratory function in crustaceans exhibiting either direct or indirect development. *Journal of Experimental Zoology* **272**, 413–418.

Spicer, J.I. & Gaston, K.J. (1997) Old and new agendas for ontogeny. *Trends in Ecology and Evolution* **12**, 381–382.

Spicer, J.I. & Morritt, D. (1996) Ontogenic changes in cardiac function in crustaceans. *Comparative Biochemistry and Physiology* **114A**, 81–89.

Spicer, J.I., Morritt, D. & Maltby, L. (1998) Effect of water-borne zinc on osmoregulation in the freshwater amphipod *Gammarus pulex* (L.) from populations that differ in their sensitivity to metal stress. *Functional Ecology* **12**, 242–247.

Spicer, J.I. & Taylor, A.C. (1987) Ionic regulation and salinity related changes in haemolymph protein in the semi-terrestrial beachflea *Orchestia gammarellus* (Pallas) (Crustacea: Amphipoda). *Comparative Biochemistry and Physiology* **88A**, 243–246.

Spicer, J.I., Thomasson, M.A. & Strömberg, J.-O. (in press) Possessing a poor anaerobic capacity does not prevent the diel vertical migration of Nordic krill *Meganyctiphanes norvegica* into hypoxic waters. *Marine Ecology—Progress Series*.

Spiers, D.E. & Baummer, S.C. (1990) Embryonic development of Japanese quail (*Coturnix coturnix japonica*) as influenced by periodic cold exposure. *Physiological Zoology* **63**, 516–535.

Spotila, J.R. (1989) Constraints of bioenergetics on animal population dynamics: an introduction to the symposium. *Physiological Zoology* **62**, 195–198.

Spradbery, J.P. & Maywald, G.F. (1992) The distribution of the European or German wasp, *Vespula germanica* (F.) (Hymenoptera: Vespidae), in Australia: past, present and future. *Australian Journal of Zoology* **40**, 495–510.

Stainier, M.W., Mount, L.E. & Bligh, J. (1984) *Energy Balance and Temperature Regulation.* Cambridge University Press, Cambridge.

Stanley, S. & Parsons, P.A. (1981) The response of the cosmopolitan species *Drosophila melanogaster*, to ecological gradients. *Proceedings of the Ecological Society of Australia* **11**, 121–130.

Staton, J.L. & Felder, D.L. (1992) Osmoregulatory capacities in disjunct western Atlantic populations of the *Sesarma reticulatum* complex (Decapoda: Grapsidae). *Journal of Crustacean Biology* **12**, 335–341.

Stauber, L.A. (1950) The problem of physiological species with special reference to oysters and oyster drills. *Ecology* **31**, 109–118.

Stave, U., ed. (1978) *Perinatal Physiology.* Plenum Press, New York.

Steffensen, J.F., Bushnell, P.G. & Schurmann, H. (1994) Oxygen consumption in four species of teleost from Greenland: No evidence for metabolic cold adaptation. *Polar Biology* **14**, 49–54.

Steinbeck, J. (1990) *The Log from the Sea of Cortez.* Mandarin, London.

Stephenson, M.J. & Knight, A.W. (1980) The effect of temperature and salinity on oxygen consumption of post-larvae of *Macrobrachium rosenbergii* (De Man) (Crustacea: Palaemonidae). *Comparative Biochemistry and Physiology* **67A**, 699–703.

Stevens, G.C. (1989) The latitudinal gradient in geographical range: How so many species coexist in the tropics. *American Naturalist* **133**, 240–256.

Stevens, C.E. & Hume, I.D. (1996) *Comparative Physiology of the Vertebrate Digestive System.* Cambridge University Press, Cambridge.

Stickle, W.B., Liu, L.L. & Foltz, D.W. (1990) Allozymic and physiological variation in populations of sea urchins (*Strongylocentrosus* spp.). *Canadian Journal of Zoology* **68**, 144–149.

Stokes, M.D. & Holland, N.D. (1996) Life-history characteristics of the Florida lancelet, *Branchiostoma floridae*: some factors affecting population dynamics in Tampa Bay. *Israel Journal of Zoology* **42** (Suppl.), 67–86.

Stone, G.N. & Willmer, P.G. (1989) Warm-up rates and body temperatures in bees: the importance of body size, thermal regime and phylogeny. *Journal of Experimental Biology* **147**, 303–328.

Storey, M. (1937) The relation between normal range and mortality of fishes due to cold at Sanibel Island, Florida. *Ecology* **19**, 10–26.

Storey, K.B. & Storey, J.M. (1992) Natural freeze tolerance in ectothermic vertebrates. *Annual Review of Physiology* **54**, 619–637.

Storey, K.B. & Storey, J.M. (1996) Natural freezing survival in animals. *Annual Review of Ecology and Systematics* **27**, 365–386.

Strathdee, A.T. & Bale, J.S. (1995) Factors limiting the distribution of *Acyrthosiphon svalbardicum* (Hemiptera: Aphididae) on Spitsbergen. *Polar Biology* **15**, 375–380.

Strohl, K.P. & Thomas, A.J. (1997) Neonatal conditioning for adult respiratory behavior. *Respiration Physiology* **110**, 269–275.

Strømme, J.A., Ngari, T. & Zachariassen, K.E. (1986) Physiological adaptations in Coleoptera on Spitsbergen. *Polar Research* **4**, 199–204.

Suarez, R.K. (1998) Oxygen and the upper limits to animal design and performance. *Journal of Experimental Biology* **201**, 1065–1072.

Sumner, F.B. & Doudoroff, P. (1938) Some experiments on temperature acclimatization and respiratory metabolism in fishes. *Biological Bulletin* **137**, 202–216.

Sutherst, R.W., Floyd, R.B. & Maywald, G.F. (1996) The potential geographic distribution of the cane toad, *Bufo marinus* L. in Australia. *Conservation Biology* **10**, 294–299.

Tait, R.V. & Dipper, F.A. (1998) *Elements of Marine Ecology.* Butterworth–Heinemann, Oxford.

Tanaka, K. (1996) Seasonal and latitudinal variation in supercooling ability of the House Spider, *Achaeranea tepidariorum* (Araneae: Theridiidae). *Functional Ecology* **10**, 185–192.

Tave, D., Smitherman, R.O. & Jayaprakas, V. (1989) Estimates of additive genetic effects, maternal effects, specific combining ability, maternal heterosis, and egg cytoplasm effects for cold tolerance in *Oreochromis niloticus* (L.). *Aquaculture and Fisheries Management* **20**, 159–166.

Taylor, W.P. (1936) What is ecology and what good is it? *Ecology* **17**, 333–346.

Taylor, E.W., Butler, P.J. & Sherlock, P.J. (1973) The respiratory and cardiovascular changes associated with the emersion response of *Carcinus maenas* (L.) during environmental hypoxia at three different temperatures. *Journal of Comparative Physiology* **86**, 95–115.

Taylor, A.C. & Spicer, J.I. (1986) Metabolic responses of the prawns *Palaemon elegans* and *P. serratus* (Crustacea: Decapoda) to acute hypoxia and anoxia. *Marine Biology* **95**, 521–530.

Taylor, A.C. & Spicer, J.I. (1988) Functional significance of a partial-emersion response in the intertidal prawn *Palaemon elegans* (Crustacea: Palaemonidae) during environmental hypoxia. *Marine Ecology—Progress Series* **44**, 141–147.

Taylor, A.C. & Spicer, J.I. (1991) Acid–base

disturbances in the haemolymph of the prawns *Palaemon elegans* (Rathke) and *P. serratus* (Pennant) (Crustacea: Decapoda). *Comparative Biochemistry and Physiology* **98A**, 445–452.

Taylor, C.R. & Weibel, E.R. (1981) Design of the mammalian respiratory system. I. Problem and strategy. *Respiration Physiology* **44**, 1–10.

Taylor, C.R., Weibel, E.R., Karas, R.H. & Hoppeler, H. (1987) Adaptive variation in the mammalian respiratory system in relation to energetic demand: VIII. Structural and functional design principles determining the limits to oxidative metabolism. *Respiration Physiology* **69**, 117–127.

Taylor, C.R., Weibel, E.R., Weber, J.M., Vock, R., Hoppeler, H., Roberts, T.J. & Brichon, G. (1996) Design of the oxygen and substrate pathways. I. Model and strategy to test symmorphosis in a network structure. *Journal of Experimental Biology* **199**, 1643–1649.

Tazawa, H., Watanabe, W. & Burggren, W.W. (1994) Embryonic heart rate in altricial birds, the pigeon (*Columba domestica*) and the bank swallow (*Riparia riparia*). *Physiological Zoology* **67**, 1448–1460.

Tedengren, M., André, C., Johannesson, K. & Kautsky, N. (1990) Genotypic and phenotypic differences between Baltic and North Sea populations of *Mytilus edulis* evaluated through reciprocal transplantations. III. Physiology. *Marine Ecology—Progress Series* **59**, 221–227.

Temple, G.K. & Johnston, I.A. (1998) Testing hypotheses concerning the phenotypic plasticity of escape performance in fish of the family Cottidae. *Journal of Experimental Biology* **201**, 317–331.

Ten Cate, G., Kooy, J.S. & Zuidweg, M.H.J. (1951) The effect of temperature on the synthesis of the enzyme cholinesterase of amphibian embryos. *Proceedings of the Royal Academy of Science, Amsterdam* **54B**, 157–170.

Terwilliger, N.B. (1998) Functional adaptations of oxygen-transport proteins. *Journal of Experimental Biology* **201**, 1085–1098.

Terwilliger, N.B. & Terwilliger, R.C. (1982) Changes in the subunit structure of *Cancer magister* hemocyanin during larval development. *Journal of Experimental Zoology* **221**, 181–191.

Terwilliger, N.B., Terwilliger, R.C. & Graham, R. (1986) Crab hemocyanin function changes during development. In: *Invertebrate Oxygen Carriers* (ed. B. Linzen), pp. 333–335. Springer-Verlag, Berlin.

Thomas, D. (1952) *Collected Poems 1934–1952.* J.M. Dent, London.

Thomas, A.J., Austin, W., Friedman, L. & Strohl, K.P. (1992) A model of ventilatory instability induced in the unrestrained rat. *Journal of Applied Physiology* **73**, 1530–1536.

Thomas, A.J., Friedman, L., MacKenzie, C.N. & Strohl, K.P. (1995) Modification of conditioned apneas in rats: Evidence for cortical involvement. *Journal of Applied Physiology* **78**, 1215–1218.

Thomas, D.W., Bosque, C. & Arends, A. (1993) Development of thermoregulation and the energetics of nestling oilbirds (*Steatornis caripensis*). *Physiological Zoology* **66**, 322–348.

Thomason, J.C., Davenport, J. & Le Comte, E. (1996) Ventilatory mechanisms and the effect of hypoxia and temperature on the embryonic lesser spotted dogfish. *Journal of Fish Biology* **49**, 965–972.

Thompson, D'A.W. (1917) *On Growth and Form.* Cambridge University Press, Cambridge.

Thompson, M.W., Merrill, D.C., Yang, G., Robillard, J.E. & Sigmund, C.D. (1995) Transgenic animals in the study of blood pressure regulation and hypertension. *American Journal of Physiology* **269**, E793–E803.

Thuesen, E.V. & Childress, J.J. (1993a) Enzymatic activities and metabolic rates of pelagic chaetognaths: lack of depth-related declines. *Limnology and Oceanography* **38**, 935–948.

Thuesen, E.V. & Childress, J.J. (1993b) Metabolic rates, enzyme activities and chemical compositions of some deep-sea pelagic worms, particularly *Nectonemertes mirabilis* (Nemertea; Hoplonemertea) and *Paeobius meseres* (Annelida; Polychaeta). *Deep-Sea Research* (Part 1), **40**, 937–951.

Thuesen, E.V. & Childress, J.J. (1994) Oxygen consumption rates and metabolic enzyme activities of oceanic California medusae in relation to body size and habitat depth. *Biological Bulletin* **187**, 84–98.

Thuesen, E.V., Miller, C.B. & Childress, J.J. (1998) Ecophysiological interpretation of oxygen consumption rates and enzymatic activities in deep-sea copepods. *Marine Ecology—Progress Series* **168**, 95–107.

Todd, C.M. (1997) Respiratory metabolism in two species of carabid beetle from the sub-Antarctic island of South Georgia. *Polar Biology* **18**, 166–171.

Tokeshi, M. (1993) Species abundance patterns and community structure. *Advances in Ecological Research* **24**, 111–186.

Toloza, E.M. & Diamond, J. (1992) Ontogenic development of nutrient transporters in rat intestine. *American Journal of Physiology* **263**, G593–G604.

Toloza, E.M., Lam, M. & Diamond, J. (1991) Nutrient extraction by cold-exposed mice: a test of digestive safety margins. *American Journal of Physiology* **261**, G608–G620.

Torgerson, C.S., Gdovin, M.J. & Remmers, J.E. (1997) Ontogeny of central chemoreception during fictive gill and lung ventilation in an *in vitro* brainstem preparation of *Rana catesbeiana. Journal of Experimental Biology* **200**, 2063–2072.

Torres, J.J. & Somero, G.N. (1988) Metabolism, enzymatic activities and cold adaptation in Antarctic mesopelagic fishes. *Marine Biology* **98**, 169–180.

Townsend, C.R. & Calow, P., eds. (1981) *Physiological Ecology: An Evolutionary Approach to Resource Use*. Blackwell Scientific Publications, Oxford.

Triplett, E.L. & Barrymore, J.D. (1960) Some aspects of osmoregulation in embryonic and adult *Cymatogaster aggregata* and other embiotocid fish. *Biological Bulletin* **118**, 472–478.

Tsuji, J.S. (1988) Seasonal profiles of standard metabolic rate of lizards (*Sceloporus occidentalis*) in relation to latitude. *Physiological Zoology* **61**, 230–240.

Turnball, D.E. & Drewes, C.D. (1996) Touch sensitivity in oligochaete giant fibres is transiently enhanced by a single spike. *Canadian Journal of Zoology* **74**, 841–844.

Tworney, A.C. (1936) Climographic studies of certain introduced and migratory birds. *Ecology* **17**, 122–132.

Tyler-Walters, H. & Davenport, J. (1990) The relationship between the distribution of genetically distinct inbred lines and upper lethal temperature in *Lasaea rubra*. *Journal of Marine Biological Association of the UK* **70**, 557–570.

Ultsch, G.R. & Cochran, B.M. (1994) Physiology of northern and southern musk turtles (*Sternotherus odoratus*) during simulated hibernation. *Physiological Zoology* **67**, 263–281.

Underwood, A.J. (1979) The ecology of intertidal gastropods. *Advances in Marine Biology* **16**, 111–210.

Unwin, D.M. (1980) *Microclimate Measurement for Ecologists*. Academic Press, London.

Ushakov, B. (1964) Thermostability of cells and proteins of poikilotherms and its significance in speciation. *Physiological Reviews* **44**, 518–560.

Väinölä, R. & Hvilsom, M.M. (1991) Genetic divergence and a hybrid zone between Baltic and North Sea *Mytilus* populations (Mytilidae: Mollusca). *Biological Journal of the Linnaean Society* **43**, 127–148.

Valentine, J.W., Erwin, D.H. & Jablonski, D. (1996) Developmental evolution of metazoan bodyplans: the fossil evidence. *Developmental Biology* **173**, 373–381.

Van Berkum, F.H. (1988) Latitudinal patterns of thermal sensitivity of sprint speed in lizards. *American Naturalist* **132**, 327–343.

Van Den Berg, C., Van Dusschoten, D., Van As, H. *et al.* (1995) Visualising the water flow in a breathing carp using NMR. *Netherlands Journal of Zoology* **45**, 338–346.

Vanhaecke, P., Siddal, S.E. & Sorgeloos, P. (1984) International study on *Artemia*. XXXII. Combined effects of temperature and salinity on the survival of *Artemia* of various geographical origins. *Journal of Experimental Marine Biological Ecology* **80**, 259–275.

Vargo, S.L. & Sastry, A.N. (1977) Acute temperature and low dissolved oxygen tolerances of brachyuran crab (*Cancer irroratus*) larvae. *Marine Biology* **40**, 165–171.

Varó, I., Taylor, A.C., Navarro, J.C. & Amat, F. (1991) Effects of temperature and oxygen tension on oxygen consumption rates of nauplii of different *Artemia* strains. *Marine Ecology—Progress Series* **76**, 25–31.

Varvio, S.-L., Koehn, R.K. & Väinölä, R. (1988) Evolutionary genetics of the *Mytilus edulis* complex in the North Atlantic region. *Marine Biology* **98**, 51–60.

Vernberg, F.J. (1959) Studies on the physiological variation between tropical and temperate-zone fiddler crabs of the genus *Uca*. III. The influence of temperature acclimation on oxygen consumption of whole organisms. *Biological Bulletin* **117**, 582–593.

Vernberg, F.J. (1962) Comparative physiology: latitudinal effects on physiological properties of animal populations. *Annual Review of Physiology* **24**, 517–546.

Vernberg, W.B. (1963) Respiration in digenetic trematodes. *Annals of the New York Academy of Sciences (Arts 1)* **113**, 261–271.

Vernberg, W.B. (1968) Physiological diversity of metabolism in marine and terrestrial Crustacea. *American Zoologist* **8**, 449–458.

Vernberg, W.B. (1969) Adaptations of host and symbionts in the intertidal zone. *American Zoologist* **9**, 357–365.

Vernberg, F.J. & Costlow, J.D. (1966) Studies on the physiological variation between tropical and temperate-zone fiddler crabs of the genus *Uca*. IV. Oxygen consumption of larvae and young crabs reared in the laboratory. *Physiological Zoology* **39**, 36–52.

Vernberg, F.J. & Vernberg, W.B. (1964) Metabolic adaptations of animals from different latitudes. *Helgolander Wissenschaftliche Meeresuntersuchungen* **9**, 476–487.

Vernberg, F.J. & Vernberg, W.B. (1966a) Studies on physiological variation between tropical and temperate zone fiddler crabs of the genus *Uca*. VII. Metabolic-temperature acclimation responses in southern hemisphere crabs. *Comparative Biochemistry and Physiology* **19**, 489–524.

Vernberg, W.B. & Vernberg, F.J. (1966b) Studies on physiological variation between tropical and temperate zone fiddler crabs of the genus *Uca*. V. Effect of temperature on tissue respiration. *Comparative Biochemistry and Physiology* **17**, 363–374.

Vernberg, W.B. & Vernberg, F.J. (1972) *Environmental Physiology of Marine Animals*. Springer-Verlag, Berlin.

Vernon, H.M. (1900) The death temperature of certain marine organisms. *Journal of Physiology* **25**, 131–136.

Village, A. (1990) *The Kestrel.* Poyser, Calton.

Vincent, I.C. & Leahy, R.A. (1997) Real-time non-invasive measurement of heart rate in working dogs: a technique with potential applications in the objective assessment of welfare problems. *Veterinary Journal* **153**, 179–183.

Visser, G.H. & Ricklefs, R.E. (1993) Development of temperature regulation in shorebirds. *Physiological Zoology* **66**, 771–792.

Vleck, C.M., Hoyt, D.F. & Vleck, D. (1979) Metabolism of avian embryos: patterns in altricial and precocial birds. *Physiological Zoology* **52**, 363–377.

Vock, R., Hoppeler, H., Claussen, H., Wu, D.X.Y., Billeter, Y., Weber, J.M., Taylor, C.R. & Weibel, E.R. (1996a) Design of the oxygen and substrate pathways. VI. Structural basis of intracellular supply to mitochondria in muscle cells. *Journal of Experimental Biology* **199**, 1689–1697.

Vock, R., Weibel, E.R., Hoppeler, H., Ørdway, G., Weber, J.M. & Taylor, C.R. (1996b) Design of the oxygen and substrate pathways. V. Structural basis of vascular substrate supply to muscle cells. *Journal of Experimental Biology* **199**, 1675–1688.

Voit, C. (1901) Über die Grösse des Energiebedarfs der Tiere im Hungerzustande. *Zeitschrift für Biology* **41**, 113–154.

Volckaert, F. & Zouros, E. (1989) Allozyme and physiological variation in the scallop *Placopecten magellanicus* and a general model for the effects of heterozygosity on fitness in marine molluscs. *Marine Biology* **103**, 51–62.

Vollestad, L.A. & Hindar, K. (1997) Developmental stability and environmental stress in *Salmo salar* (Atlantic salmon). *Heredity* **78**, 215–222.

Voogt, P.A., Broertjes, J.J.S. & Oudejans, R.C.H.M. (1985) Vitellogenesis in sea star: physiological and metabolic implications. *Comparative Biochemistry and Physiology* **80A**, 141–147.

Wache, S., Terwilliger, N.B. & Terwilliger, R.C. (1988) Hemocyanin structure changes during development of the crab *Cancer productus*. *Journal of Experimental Zoology* **247**, 23–32.

Waddington, C.H. (1959) Canalization of development and genetic assimilation of acquired characters. *Nature* **183**, 1654–1655.

Walker, D.G. (1965) Development of hepatic enzymes for the phosphorylation of glucose and fructose. *Advances in Enzyme Regulation* **3**, 163–184.

Walker, P.A. (1990) Modelling wildlife distributions using a geographic information system: kangaroos in relation to climate. *Journal of Biogeography* **17**, 279–289.

Walsberg, G.E. & Wolf, B.O. (1996) An appraisal of operative temperature mounts as tools for studies of ecological energetics. *Physiological Zoology* **69**, 658–681.

Walsh, S.J., Haney, D.C., Timmerman, C.M. & Dorazio, R.M. (1998) Physiological tolerances of juvenile robust redhorse, *Moxostoma robustum*: conservation implications for an imperilled species. *Environmental Biology of Fishes* **51**, 429–444.

Walsh, W.A., Swanson, C. & Lee, C.-S. (1991a) Combined effects of temperature and salinity on embryonic development and hatching of striped mullet, *Mugil cephalus*. *Aquaculture* **97**, 281–290.

Walsh, W.A., Swanson, C. & Lee, C.-S. (1991b) Effects of development, temperature and salinity on metabolism in eggs and yolk-sac larvae of milkfish, *Chanos chanos* (Forskal). *Journal of Fish Biology* **39**, 115–125.

Walsh, W.A., Swanson, C., Lee, C.-S., Banno, J.E. & Eda, H. (1989) Oxygen consumption by eggs and larvae of striped mullet, *Mugil cephalus*, in relation to development, salinity and temperature. *Journal of Fish Biology* **35**, 347–358.

Walton, M. (1988) Relationships among metabolic, locomotory, and field measures of organismal performance in the Fowler's toad (*Bufo woodhousii fowleri*). *Physiological Zoology* **61**, 107–118.

Walton, M. (1993) Physiology and phylogeny: the evolution of locomotor energetics in hylid frogs. *American Naturalist* **141**, 26–50.

Wan, Q., Liao, M., Brown, G.M. & Pang, S.F. (1997) The development and circadian variation of melatonin receptors in the chicken spinal cord. *Developmental Neuroscience* **19**, 196–201.

Wang, T., Burggren, W.W. & Nobrega, E. (1995) Metabolic, ventilatory, and acid-base responses associated with specific dynamic action in the toad *Bufo marinus*. *Physiological Zoology* **68**, 192–205.

Wang, R., Zhang, P., Gong, Z. & Hew, C.L. (1995) Expression of the antifreeze protein gene in transgenic goldfish (*Carassius auratus*) and its implication in cold adaptation. *Molecular Marine Biology and Biotechnology* **4**, 20–26.

Ward, S. (1996) Energy expenditure of female barn swallows *Hirundo rustica* during egg formation. *Physiological Zoology* **69**, 930–951.

Ward, R., Blandon, I.R., King, T.L. & Beitinger, T.L. (1993) Comparisons of critical thermal maxima and minima of juvenile red drum (*Sciaenops ocellatus*) from Texas and North Carolina. *Northeast Gulf Science* **13**, 23–28.

Ward, D. & Seely, M.K. (1996) Adaptation and constraint in the evolution of the physiology and behavior of the Namib Desert tenebrionid beetle genus *Onymacris*. *Evolution* **50**, 1231–1240.

Ward, M.P., Milledge, J.S. & West, J.B. (1995) *High Altitude Medicine and Physiology*, 2nd edn. Chapman & Hall Medical, London.

Waterhouse, F.L. (1950) Humidity and temperature in grass microclimates with reference to insulation. *Nature* **166**, 232–233.

Watkins, T.B. (1996) Predator-mediated selection on burst swimming performance in tadpoles of the Pacific tree frog, *Pseudacris regilla*. *Physiological Zoology* **69**, 154–167.

Watt, W.B. (1985) Bioenergetics and evolutionary genetics: opportunities for a new synthesis. *American Naturalist* **125**, 118–143.

Weatherhead, P.J. (1986) How unusual are unusual events? *American Naturalist* **128**, 150–154.

Weathers, W.W. & Sullivan, K.A. (1993) Seasonal patterns of time and energy allocation by birds. *Physiological Zoology* **66**, 511–536.

Webb, D.R. & McClure, P.A. (1989) Development of heat production in altricial and precocial rodents: implications for the energy allocation hypothesis. *Physiological Zoology* **62**, 1293–1315.

Weber, R.E. (1995) Hemoglobin adaptations to hypoxia and altitude: the phylogenetic perspective. In: *Hypoxia and the Brain* (eds J.R. Sutton, C.S. Houston & G. Coates), pp. 31–44. Queen City Printers, Burlington, Vermont.

Weber, J.M., Brichon, G., Zwingelstein, G. *et al.* (1996a) Design of the oxygen and substrate pathways. IV. Partitioning energy provision from fatty acids. *Journal of Experimental Biology* **199**, 1667–1674.

Weber, J.M., Roberts, T.J., Vock, R., Weibel, E.R. & Taylor, C.R. (1996b) Design of the oxygen and substrate pathways. III. Partitioning energy provision from carbohydrates. *Journal of Experimental Biology* **199**, 1659–1666.

Wehner, R., Marsh, A.C. & Wehner, S. (1992) Desert ants on a thermal tightrope. *Nature* **357**, 586–587.

Weibel, E.R., Taylor, C.R. & Bolis, L.C. (1998) *Principles of Animal Design. The Optimisation and Symmorphosis Debate*. Cambridge University Press, Cambridge.

Weibel, E.R., Taylor, C.R. & Hoppeler, H. (1991) The concept of symmorphosis: a testable hypothesis of structure–function relationship. *Proceedings of the National Academy of Sciences of the USA* **88**, 10357–10361.

Weibel, E.R., Taylor, C.R. & Hoppeler, H. (1992) Variations in function and design: testing symmorphosis in the

respiratory system. *Respiration Physiology* **87**, 325–348.

Weibel, E.R., Taylor, C.R., Hoppeler, H. & Karas, R.H. (1987) Adaptive variation in the mammalian respiratory system in relation to energetic demand: I. Introduction to problem and strategy. *Respiration Physiology* **69**, 1–6.

Weibel, E.R., Taylor, C.R., Weber, J.M. *et al.* (1996) Design of the oxygen and substrate pathways. VII. Different structural limits for oxygen and substrate supply to muscle mitochondria. *Journal of Experimental Biology* **199**, 1699–1709.

Weithe, W.H., ed. (1964) *The Physiological Effects of High Altitude*. Pergamon Press, London.

Wells, R.M.G. (1987) Respiration of Antarctic fish from McMurdo Sound. *Comparative Biochemistry and Physiology* **88A**, 417–424.

Wells, J.V. & Richmond, M.E. (1995) Populations, metapopulations, and species populations—what are they and who should care? *Wildlife Society of Bulletin* **23**, 458–462.

Wernig, A., Irintchev, A. & Weisshaupt, P. (1990) Muscle injury, cross-sectional area and fibre type distribution in mouse soleus after intermittent wheel-running. *Journal of Physiology* **428**, 639–652.

Werntz, H.O. (1963) Osmotic regulation in marine and fresh-water gammarids (Amphipoda). *Biological Bulletin* **124**, 225–239.

Wesolowski, T. (1994) Variation in the numbers of resident birds in a primaeval temperate forest: are winter weather, seed crop, caterpillars and interspecific competition involved? In: *Bird Numbers 1992: Distribution, Monitoring and Ecological Aspects* (eds E.J.M. Hagemeijer & T.J. Verstrael), pp. 203–211. Statistics Netherlands, Voorburg/Heerlen & SOVON, Beek-Ubbergen, The Netherlands.

West, J.B. (1998) *High Life: A History of High-Altitude Physiology and Medicine*. Oxford University Press, New York.

West-Eberhard, M.J. (1989) Phenotypic plasticity and the origins of diversity. *Annual Review of Ecology Systematics* **20**, 249–278.

von Westernhagen, H. (1988) Sublethal effects of pollutants on fish eggs and larvae. In: *Fish Physiology*, Vol XIA (eds W.S. Hoar & D.J. Randall), pp. 253–346. Academic Press, New York.

von Westernhagen, H. & Dethlefsen, V. (1975) Combined effects of cadmium and salinity on development and survival of flounder eggs. *Journal of the Marine Biological Association of the UK* **55**, 945–957.

von Westernhagen, H., Dethlefsen, V. & Rosenthal, H. (1979) Combined effects of cadmium, copper and lead on developing herring eggs and larvae. *Helgolander*

Wissenschaftliche Meeresunteruchungen **32**, 257–278.

Wethey, D.S. (1985) Catastrophe, extinction, and species diversity: a rocky intertidal example. *Ecology* **66**, 445–456.

Whitehead, P.J., Webb, G.J.W. & Seymour, R.S. (1990) Effect of incubation temperature on development of *Crocodylus johnstoni* embryos. *Physiological Zoology* **63**, 949–964.

Whiteley, N.M., Taylor, E.W. & El Haj, A.J. (1996) A comparison of the metabolic cost of protein synthesis in stenothermal and eurythermal isopod crustaceans. *American Journal of Physiology* **271**, R1295–R1303.

Whitlock, M. (1996) The heritability of fluctuating asymmetry and the genetic control of developmental stability. *Proceedings of the Royal Society of London* **263B**, 849–853.

Whitney, R.J. (1939) The thermal resistance of mayfly nymphs from ponds and streams. *Journal of Experimental Biology* **16**, 374–385.

Whittaker, J.B. & Tribe, N.P. (1996) An altitudinal transect as an indicator of responses of a spittlebug (Auchenorrhyncha: Cercopidae) to climate change. *European Journal of Entomology* **93**, 319–324.

Whittow, G.C. & Tazawa, H. (1991) The early development of thermoregulation in birds. *Physiological Zoology* **64**, 1371–1390.

Widdows, J. (1985) The effects of fluctuating and abrupt changes in salinity on the performance of *Mytilus edulis*. In: *Marine Biology of Polar Regions and Effects of Stress on Marine Organisms* (eds J.S. Gray & M.E. Christiansen), pp. 555–566. John Wiley, Chichester.

Widdows, J., Donkin, P., Salkeld, P.N., Cleary, J.J., Lowe, D.M., Evans, S.V. & Thomson, P.E. (1984) Relative importance of environmental factors in determining physiological differences between two populations of mussels (*Mytilus edulis*). *Marine Ecology—Progress Series* **17**, 33–47.

Wiesel, T.N. (1982) Postnatal development of the visual cortex and the influence of the environment. *Nature* **299**, 583–591.

Wieser, W. (1984) A distinction must be made between the ontogeny and the phylogeny of metabolism in order to understand the mass exponent of energy metabolism. *Respiration Physiology* **55**, 1–9.

Wigly, T.M.L. (1985) Impact of extreme events. *Nature* **316**, 106–107.

Williams, P.H. & Humphries, C.J. (1996) Comparing character diversity among biotas. In: *Biodiversity: A Biology of Numbers and Difference* (ed. K.J. Gaston), pp. 54–76. Blackwell Science, Oxford.

Williams, D.D. & Williams, N.E. (1998) Aquatic insects in an estuarine environment: densities, distribution and salinity tolerance. *Freshwater Biology* **39**, 411–421.

Williamson, M. (1996) *Biological Invasions*. Chapman & Hall, London.

Willmer, P.G. (1980) The effects of a fluctuating environment on the water relations of larval Lepidoptera. *Ecological Entomology* **5**, 271–292.

Willmer, P.G. (1981) Microclimate and the environmental physiology of insects. *Advances in Insect Physiology* **16**, 1–57.

Wilson, R.P. & Grémillet, D. (1996) Body temperatures of free-living African penguins (*Spheniscus demersus*) and Bank cormorants (*Phalacrocorax neglectus*). *Journal of Experimental Biology* **199**, 2215–2223.

Withers, P.C. (1992) *Comparative Animal Physiology*. Saunders College, Philadelphia.

Woakes, A.J. & Foster, W.A., eds. (1991) The comparative physiology of exercise. *Journal of Experimental Biology* **160**, 1–340.

Wohlschlag, D.E. (1960) Metabolism of an Antarctic fish and the phenomenon of cold adaptation. *Ecology* **41**, 287–292.

Wohlschlag, D.E. (1962) Metabolic requirements for the swimming activity of three Antarctic fishes. *Science* **137**, 1050–1051.

Wolf, B.O., Wooden, K.M. & Walsberg, G.E. (1996) The use of thermal refugia by two small desert birds. *Condor* **98**, 424–428.

Wolf, C.M., Griffith, B., Reed, C. & Temple, S.A. (1996) Avian and mammalian translocations: update and reanalysis of 1987 survey data. *Conservation Biology* **10**, 1142–1154.

Wood, C.M. & McDonald, D.G., eds. (1996) *Global Warming: Implications for Freshwater and Marine Fish*. Cambridge University Press, Cambridge.

Woods, W.A. Jr & Stevenson, R.D. (1996) Time and energy costs of copulation for the sphinx moth, *Manduca sexta*. *Physiological Zoology* **69**, 682–690.

Wright, P.A. (1995) Nitrogen excretion: three end products, many physiological roles. *Journal of Experimental Biology* **198**, 273–281.

Wright, W.G., Kirscman, D., Rozen, D. & Maynard, B. (1996) Phylogenetic analysis of learning-related neuromodulation in molluscan mechanosensory neurons. *Evolution* **50**, 2248–2263.

Wright, P.M. & Wright, P.A. (1996) Nitrogen metabolism and excretion in bullfrog (*Rana catesbeiana*) tadpoles and adults exposed to elevated environmental ammonia levels. *Physiological Zoology* **69**, 1057–1078.

Yahav, S., Simson, S. & Nevo, E. (1988a) Adaptive energy metabolism in four chromosomal species of subterranean mole rats. *Oecologia* **77**, 533–536.

Yahav, S., Simson, S. & Nevo, E. (1988b) Total body water and adaptive turnover rate in four chromosomal mole rat species of the *Spalax ehrenbergi* superspecies in Israel. *Journal of Zoology* **218**, 461–470.

Yalden, D.W. & Pearce-Higgins, J.W. (1997) Density-dependence and winter weather as factors affecting the size of a population of golden plovers *Pluvialis apicaria*. *Bird Study* **44**, 227–234.

Yayanos, A.A. (1978) Recovery and maintenance of live amphipods at a pressure of 580 bars from an ocean depth of 5700 meters. *Science* **200**, 1056–1059.

Yayanos, A.A. (1981) Reversible inactivation of deep-sea amphipods (*Paralicella capresca*) by a decompression of 601 bars from atmospheric pressure. *Comparative Biochemistry and Physiology* **69A**, 563–565.

Young, A. (1997) Ageing and physiological functions. *Philosophical Transactions of the Royal Society of London* **352B**, 1837–1843.

Young, A.M. (1991) Temperature-salinity tolerance of two latitudinally-separated populations of the long-wrist hermit crab *Pagurus longicarpus* (Crustacea, Decapoda, Paguridae). *Ophelia* **34**, 29–40.

Young, R.T. (1941) The distribution of the mussel (*Mytilus californianus*) in relation to the salinity of its environment. *Ecology* **22**, 379–386.

Young, S.R. (1979) Respiratory metabolism of *Alaskozetes antarcticus*. *Journal of Insect Physiology* **25**, 361–369.

Young, T.P. (1994) Natural die-offs of large mammals: implications for conservation. *Conservation Biology* **8**, 410–418.

Young, C.M. & Tyler, P.A. (1993) Embryos of the deep-sea echinoid *Echinus affinis* require high pressure for development. *Limnological*

Oceanography **38**, 178–181.

Young, C.M., Tyler, P.A. & Emson, R.H. (1996a) Embryonic pressure tolerances of bathyal and littoral echinoids from the tropical Atlantic and Pacific oceans. In: *Echinoderm Research 1995 Proceedings of the 4th European Echinoderms Colloquium* (eds R.H. Emson, A.B. Smith, A.B. & A.C. Campbell). A.A. Balkema, Rotterdam.

Young, C.M., Tyler, P.A. & Gage, J.D. (1996b) Vertical distribution correlates with pressure tolerances of early embryos in the deep sea asteroid *Plutonaster bifrons*. *Journal of the Marine Biological Association of the UK* **76**, 749–757.

Young-Lai, W.W., Charmantier-Daures, M. & Charmantier, G. (1991) Effect of ammonia on survival and osmoregulation in different life stages of the lobster *Homarus americanus*. *Marine Biology* **110**, 293–300.

Zamar, W.E. & Mangum, C.P. (1979) Irreversible non-genetic temperature adaptation of oxygen consumption in clones of the sea anemone (Verrill). *Biological Bulletin* **157**, 536–547.

Zamudio, K.R., Huey, R.B. & Crill, W.D. (1995) Bigger isn't always better: body size, developmental and parental temperature and male territorial success in *Drosophila melanogaster*. *Animal Behavior* **49**, 671–677.

Zar, J.H. (1996) *Biostatistical Analysis*, 3rd edn. Prentice-Hall, London.

Zwaan, B.J., Bijlsma, R. & Hoekstra, R.F. (1992) On the developmental theory of aging. II. The effect of developmental temperature on longevity in relation to adult body size in *D. melanogaster. Heredity* **68**, 123–130.

Index